WIND

WIND

HOW THE FLOW OF AIR HAS SHAPED LIFE, MYTH, AND THE LAND

JAN DEBLIEU

A MARINER BOOK
HOUGHTON MIFFLIN COMPANY
BOSTON · NEW YORK

First Mariner Books edition 1999

Library of Congress Cataloging-in-Publication Data
DeBlieu, Jan.
Wind : how the flow of air has shaped life, myth, and the land / Jan DeBlieu.
p. cm.
Includes bibliographical references and index.
ISBN 0-395-78033-0
ISBN 0-395-95794-x (pbk.)
1. Winds. I. Title.
QC931.D4 1998
551.51'8 — dc21 98-16851 CIP

Printed in the United States of America

Book design by Robert Overholtzer

QUM 10 9 8 7 6 5 4 3 2 1

Portions of Chapter 1 were published in the magazines *Outer Banks* and *Orion;* portions of Chapter 4 appeared in *Audubon;* and portions of Chapter 9 in *Air & Space/Smithsonian.*

FOR MOM AND DAD

with gratitude and love

ACKNOWLEDGMENTS My debts are many and varied. In the course of my research, dozens of people told me of ways in which the wind had touched their lives and the lives of people they know. Without those nuggets, this book would have been much poorer in material and narrower in scope. Thanks to them all.

The scientists named within these pages were all tremendously helpful. Three gave so generously of their time that they deserve special mention: Karen Havholm of the University of Wisconsin at Eau Claire, who is a consummate teacher; Bob Rice of Weather Window in Wolfeboro, New Hampshire; and Steven Vogel of Duke University in Durham, North Carolina. I would also like to thank the staff of the Manteo branch of the Dare County Library, especially Veronica McMurran Brickhouse.

My idea for a book about the wind was amorphous, at best, when Lizzie Grossman, then of Sterling Lord Literistic, gave me the encouragement I needed to begin working on it full-time. I am grateful to her. At Houghton Mifflin my editor, Harry Foster, and my manuscript editor, Peg Anderson, did a masterful job of holding my feet to the fire (gently).

The task of writing a book can be as much an emotional journey as an intellectual one. For both their constant support and their nuts-and-bolts assistance with various problems I would like to thank Steve Brumfield, Marcia Lyons, Nancy Cowal, Shay Clanton, Beth and Patrick Martin, Janet Walsh, Sharon LaPalme, Jim Fineman, Maria Merritt, and Stuart Chaitkin.

One warm spring day as we sat on the beach, my husband, Jeff, said rather offhandedly, "Why don't you write a book about the wind?" With that simple question he mapped the direction of my life for the next several years. I am indebted to him for the book's conception and for the long period of my life that he has served as my mentor. Reid, our son, graciously shared me with the work he knows I love. My parents, Helen and Ivan DeBlieu, offered unflagging love and support. These last four are the pillars of my world; they are my cardinal winds.

No one can tell me,
    Nobody knows,
Where the wind comes from,
    Where the wind goes.

— *A. A. Milne*

. . . Come like the wind and cleanse.

— *Episcopal prayer*

Contents

ONE

Into the Dragon's Mouth

IT BEGINS with a subtle stirring caused by sunlight falling on the vapors that swaddle the earth. It is fueled by extremes — the stifling warmth of the tropics, the bitter chill of the poles. Temperature changes set the system in motion: hot air drifts upward and, as it cools, slowly descends. Knots of high and low pressure gather strength or diminish, forming invisible peaks and valleys in the gaseous soup.

Gradually the vapors begin to swirl as if trapped in a simmering cauldron. Air molecules are caught by suction and sent flying. They slide across mountain ridges and begin the steep downward descent toward the barometric lows. As the world spins, it brushes them to one side but does not slow them.

Tumbling together, the particles of air become a huge, unstoppable current. Some of them rake the earth, tousling grasses and trees, slamming into mountains, pounding anything that stands in their way. They are a force unto themselves, a force that shapes the terrestrial and aquatic world. They bring us breath and hardship. They have become the wind.

I STAND on a beach near sunset, squinting into the dragon's mouth of a gale. The wind pushes tears from the corners of my eyes across my temples. Ocean waves crest and break quickly, rolling onto the beach like tanks, churned to an ugly, frothy blue-brown. The storm is a typical northeaster, most common in spring but also likely to occur in January or June.

Where I live, on North Carolina's Outer Banks, the days are defined by wind. Without it the roar of the surf would fall silent; the ocean would become as languid as a lake. Trees would sprout wherever their seeds happened to fall, cresting the frontal dune, pushing a hundred feet up with spreading crowns. We would go about our lives in a vacuum. That is how it feels in the few moments when the wind dies: ominous, apocalyptic. As if the world has stopped turning.

I lounge on the beach with friends, enjoying a mild afternoon. A light west breeze lulls and then freshens from the east. Its salty tongue is cooling and delightful at first, but as the gusts build to fifteen miles an hour we begin to think of seeking cover. We linger awhile — how long can we hold out, really? — until grains of sand sting our cheeks and fly into our mouths. As we climb the dune that separates us from the parking lot, I am struck anew by the squatness of the landscape. Nothing within a half-mile of the ocean grows much higher than the dune line. Nothing can withstand the constant burning inflicted by maritime wind.

On this thread of soil that arches twenty miles east of the mainland, every tree and shrub must be adapted to living in wind laden with salt and ferociously strong. Gusts of fifty miles an hour or more will shatter any limbs that are less pliant than rubber. We have no protection from raw weather here; we are too far out to sea. There is nothing between the coast and the Appalachian Mountains, hundreds of miles inland, to brake the speed of building westerly breezes. There is nothing between the Outer Banks and Africa to dampen the force of easterly blows.

Any book on weather will tell you that winds are caused by the uneven heating of the earth. Pockets of warm and cold air jostle each other, create an airflow, and *voilà!* the wind begins to blow. Air moves from high pressure to low pressure, deflected to the right or the left by the rotation of the earth. It is a simple matter of physics. I try to keep that fact in mind as I stand on the beach, bent beneath the sheer force of air being thrown at me, my hair beating against my eyes. Somehow, out in the elements, the wisdom of science falls a bit short. It is easier to believe that wind is the roaring breath of a serpent who lives just over the horizon.

THE WIND, the wind. It has nearly as many names as moods: there are siroccos, Santa Anas, foehns, brickfielders, boras, williwaws, chinooks, monsoons. It has, as well, unrivaled power to evoke comfort or suffering, bliss or despair, to bless with fortune, to tear apart empires, to alter lives. Few other forces have so universally shaped the diverse terrains and waters of the earth or the plants and animals scattered through them. Few other phenomena have exerted such profound influence on the history and psyche of humankind.

From the soft stirrings that rustle leaves and grasses on summer afternoons to the biting storms that threaten life and limb, wind touches us all every day of our lives. We pay homage to its presence or absence each time we dress to go outside. We worship it with sighs, curses, and tears. "She's blowin', she is," the captain of a commercial fishing boat told me one stormy day shortly after I moved to the Outer Banks. As I struggled to keep my footing on a salt-slicked dock, I had to agree. Thereafter I made the expression part of our household vernacular. "She's blowin', she is," my husband and I joked those first winters, as Arctic-born breezes set our teeth on edge. She is, she is. But what in God's name *is* she?

In strict scientific terms wind is scarcely more than a clockwork made up of gaseous components. The heat of the sun and the rotation of the earth set the system ticking and keep it wound. The gears are simply air's inherent tendency to rise when heated and fall when cooled.

These patterned movements of air fasten into place the bands of wind and calm that girdle our small globe. A belt of constant low pressure rings the earth's middle, a weather equator that creates a strip of general breezelessness popularly known as the doldrums. To the north and the south the moist breath of the trade winds stirs the atmospheric stew. The pleasant trade regions are bounded in the northern and southern hemispheres by the comparatively stagnant zones known as the horse latitudes — so named, legend holds, because calm air within them slowed the sailing ships of early explorers and forced crew members to conserve water by throwing horses overboard. A significant portion of the world's deserts lie within the horse latitudes. Above and below 35 degrees, in the two breezy zones that encompass most of North America, Europe, China, Argentina, Chile, and New Zealand, prevailing westerlies revive the flow. These give way to bands of light easterlies that encircle the farthest, coldest reaches of the earth.

The atmosphere's alternating punches are felt most solidly in the southern hemisphere, where large expanses of open ocean enable winds to gather serious power. In the northern hemisphere the major continents harbor standing cells of high pressure; the wind must weave its way between pressure cores and over land features — mountains, valleys, cities — that muddy its flow and retard its passage. But in the north, wind unleashes catastrophic strength in the form of tornadoes that shred entire towns and northeasters that set seas chopping at shorelines like ravenous beasts.

The two contrasting faces of wind — its predictability and its moodiness — imbue it with the qualities of an animate being. Like the human body, wind is much more than the simple sum of its parts. Cool, gentle breezes seem peculiarly designed to nurture and heal, while storms strike us as personifications of the wrath of God. "This is the disintegrating power of a great wind: it isolates one from one's kind," Joseph Conrad wrote in *Typhoon.* "An earthquake, a landslip, an avalanche, overtake a man incidentally, as it were — without passion. A furious gale attacks him like a personal enemy, tries to grasp his limbs, fastens upon his mind, seeks to rout his very spirit out of him."

Out in a tempest on a boat being tossed by angry seas, it is difficult to think of the wind as a passive player in deciding one's fate. Wind has served as the pivotal factor in many lives, and in the histories of many peoples. Early explorers, from the Polynesians to the Vikings to the Spaniards, were led by favorable breezes to follow certain routes. As a result, Brazil was settled by Europeans more than a century before the west coast of Africa was, and Cortés ravaged the Aztecs and dismantled their empire a hundred years before the Pilgrims landed at Plymouth Rock. Key battles have been won and entire armies vanquished because of fortuitous turns of wind.

I like to speculate about how the world might be different if wind had arranged itself in other patterns, defying physics. Which would be the richer nations and which the poorer? Where might the rain forests lie and the great deserts? What of the history of this country? In 1777 George Washington defeated Charles Cornwallis in a crucial battle that turned when a north wind froze muddy roads along the Delaware River and enabled the new Americans to quickly reposition their artillery. If not for that wind, might we still be subjects of a distant queen or king?

BETWEEN 35 and 36 degrees north latitude, the thin islands known as the Outer Banks lie in a band of spirited west wind that accelerates as it moves over the Piedmont region and toward the Atlantic Ocean. The weather of this coast is shaped by the westerlies that scream across the continent in winter, pushing calmer, milder air far south.

Offshore the wide, warm Gulf Stream ropes its way north past Cape Hatteras and turns back out to sea after a close swipe at land. It mingles briefly with the cold, dying tongues of the southbound Labrador Current. In terms of weather, the junction of these two flows is enough to stop the show. In winter, when a dome of high pressure from the Arctic drifts southeast, it may come to the edge of the Gulf Stream and stall.

Will it linger or be pushed over the Gulf Stream and out to sea? Suppose there is a core of warm air off the coast, just east of the Stream. At the same time, suppose the jet stream has grown unusually strong and is flowing toward the northeast. The two air masses bump against each other like huge bubbles, the cold air fighting to move east, the warm air prodded north by the jet stream. A pocket of turbulence develops in the crook between them. Wind flows east, then is bent quickly to the north. Unable to resist the centrifugal force, it begins to move full circle, creating a system of low pressure that deepens violently.

The barometer plummets; rain descends in torrents. Up north, snow falls thick and fast. The western edge of the Gulf Stream is where great winter storms are made. They drift north, bequeathing rain to the Outer Banks usually, but sometimes snow. And wind.

In the spring of 1962 an explosive low-pressure system developed unexpectedly over the Outer Banks. In the wake of fierce northeast winds the ocean pounded the shore for three days, spilling over the dunes and through the little towns tucked behind them. During that particular meteorological episode, known as the Ash Wednesday Storm, people woke to find the ocean sloshing into their beds. This cycle of weather has been repeated many times since, though never with a force equal to that of the first.

Such sudden, lashing northeasters have always intrigued coastal forecasters, who as recently as the early 1980s were at a loss to explain them. Now, with the help of Doppler radar, satellite photographs, and computer models of the atmosphere, meteorologists can often tell when a winter low-pressure system threatens to form over the coast. They can warn island residents, with some confidence, to buckle down for a storm.

More typically the wind blows fickle, and its swings of mood are devilishly tricky to foretell. At the center of a pressure core the wind speed slows, but at the edges it quickens. A strong knot of high pressure, sliding over the coast, may bring light wind that lasts for days. The system may stall long enough to dissolve, or it may venture out to sea, stirring up gales as it passes.

How much wind tomorrow? Technicians at the weather station make their educated guesses, knowing all along that the wind may fool them. Knowing that whatever else it does, the wind will call the day's tune.

*

BEFORE THE ADVENT of worldwide forecasting systems, islanders watched for subtle changes to predict the behavior of weather and wind. They studied the sky and the animals the way a mother might look for the telltale signs that her young child is growing tired and cross. If, in a light, variable wind the gulls stand facing north, watch for steady north wind by nightfall. If clouds form a halo around the moon, count the stars within the halo. If there are three, expect bad weather for the next three days.

A mackerel sky — one with clouds that look like fish scales — means rain is on the way. A sundog at sunset foretells a bad storm. A mild spell in December or January is a "weather breeder"; it brings penetrating cold before winter's end. "A warm Christmas," an elderly island man once told me, "makes a fat cemetery."

The intensity of the weather here always depends on the wind, and the traditional sayings impart more folk wisdom about gales and breezes than about any other facet of life. A heavy dew in the morning means heavy wind by afternoon. If a swarm of biting flies shows up on a fishing boat far offshore, a land breeze is bound to shift to an ocean breeze. When the wind swings hard to the northeast, it will most likely blow itself out in a day:

> A Saturday shift, come late or soon,
> It seldom stands till Sunday noon.

Once or twice a winter, however, a northeaster lasts for most of a week. No matter how it begins or ends, local lore holds that the blow will always diminish on the third, or fifth, or seventh day, never on the second, fourth, or sixth day.

Only fools lived at the edge of the ocean back before hurricanes could be spotted on radar. The houses of Outer Banks natives nestled together in wooded sections just off Albemarle and Pamlico sounds. The sound side was considered the front of the islands, and the ocean beach, where the fury of storms hit hardest, was thought of as the back. It was the jumping-off point, the place where daring souls — swimmers, sailors, fishermen — could venture from the encircling arms of a continent into an ocean of uncertainty and terror.

Islanders spoke of their homeland as if they were intent on keeping their backs to the wind.

The cattle that ranged freely across the Outer Banks in the late nineteenth and the early twentieth century seemed to know when a weather shift was imminent, and they anticipated changes in the wind to escape biting flies. If they moved to the "back of the beach," east wind was on the way. If they migrated to the marshes, the easterly breeze would swing west. Most of the time the range stock stayed in the open grasslands and dunes. When they wandered into the villages, residents began boarding up windows for a hurricane.

Normally the wind migrates slowly from northeast to east to southeast to southwest, moving clockwise in the anticyclonic pattern typical of high-pressure systems in the northern hemisphere. There are exceptions, of course, when the wind direction shifts backward — counterclockwise. For generations native islanders have known such a pattern to be a harbinger of the most violent storms. The weather change might come as a localized thunderstorm or a devastating hurricane, but a backing wind is always to be feared. As an old saying has it, "I'd rather look at Grandma's drawers than see a backing wind."

WIND IS culture and heritage on the Outer Banks; wind shapes earth, plant, animal, human. Wind toughens us, moves mountains of sand as we watch, makes it difficult to sleepwalk through life.

The spring I moved to the islands I lived in a house beset by wind. Air seeped easily through the decayed siding and whistled through the roof. The constant clatter made me lonely and chafed my nerves, but I gladly sought the shelter of those rooms rather than stand exposed to the chilling breeze. I developed a ritual for going out: before opening the door I pulled on my coat and gloves, yanked down my hat, and braced myself for an onslaught.

I conditioned myself slowly, taking walks in steady wind for twenty minutes at first, with the hope of working up to forty-five. An appreciation for wind was not in my nature; I had to learn to like the feel of air pummeling my chest and roaring across my skin.

"Light" wind, I learned, blew less than fifteen miles an hour.* Anything less than ten miles an hour was not worthy of mention.

Walking with my hood pulled hard against my scalp, I began to notice how animals coped with wind. Terns, the kamikazes of the bird world, seemed oblivious even to hard gales. I remember watching them one spring afternoon at Oregon Inlet, as air howled down on us from the north and waves sloshed against each other. Together wind and tide made a mess of the landscape; with the frothing water and the whipping branches of wax myrtle shrubs, it seemed as if the world were being shaken to its foundations. Yet the terns hung steady in midair, flapping their wings quickly and chittering to each other, their beaks pointed downward as they scanned the ocean for fish.

Not many animals come out in such wind. Those that do may find the normal parameters of life redrawn. In a sustained east wind the water in the sounds is pushed toward the mainland so that vast stretches of sandy bottom are exposed. Islanders refer to this as the tide running out, and indeed it is the only kind of falling tide to be seen on the banks' western shore. The water level in the estuaries here does not respond to the pull of the moon. All sound tides are erratic and driven strictly by wind; they ebb in northeasters and flow during westerlies.

Soon after I moved here I learned that water swept east by wind for many miles has a way of suddenly spilling over its normal banks, like a bowl tipped sloppily to one side. One morning after several days of hard west wind, I parked in a lot near a fish house on Pamlico Sound. An islander casually warned me, "You might ought to move your car, case we get some tide." I parked on higher ground. Within an hour three feet of briny water covered the fish-house lot.

Even the more docile winds affect the shape of the water and the distribution of creatures within it. East winds send the surf pounding against the beach; west winds slow the shoreward roll of breakers

*All wind speeds are given here in miles per hour rather than metric or nautical measurements. At fifteen miles an hour, waves begin to form whitecaps; at thirty-two, a breeze becomes a gale; at seventy-four, a tropical storm is upgraded to a hurricane.

and make them stand erect. The best surfing waves are sculpted by a northeast blow that shifts cleanly to the west. But if the west wind blows too long, the breakers are knocked flat. Surfers disappear, replaced by commercial fishermen, who row dories just offshore to set their nets for bluefish and trout.

We all have our favorite winds. Outer Banks surf casters like a land breeze because, as they say,

> Wind from the east, fish bite the least.
> Wind from the west, fish bite the best.

A westerly breeze draws trout, mullet, and other species to the calm waters in the lee of the shore. During duck hunting season it also pushes waterfowl from the middle of Pamlico Sound toward the islands, putting them in easy range of hunting blinds. A friend of mine, an avid hunter and fisherman who lives on Hatteras Island, grew so enamored of the sound-side breeze that he vowed to name his first-born son West Wind. His wife's wisdom prevailed; they named the child Teal.

Good fishing or poor, the light summer easterlies are dearest to my heart. West winds muddy the ocean waters, but east winds clear them. West winds bring biting flies to the beach, but east winds banish them to the marsh. The most pleasant summer days are those with an ocean breeze strong enough to set up a little surf but not so strong as to make swimming dangerous. Waves roll lazily ashore as wind gently fills my lungs, caresses my skin, and sweeps cobwebs from my brain. I lie in the sun, hot but cool enough for reading. I slip noiselessly into the clear green surf and float on top, watching as sparkling grains of sand tumble out to sea between waves.

I LIVE in an island forest now, where tree trunks slash the winter wind before it can hit the house full force. At night I listen to the loblolly pines pitching back and forth high overhead and wonder how many more years the cottages on the ocean will be able to stand against the forces that batter them.

At times I imagine that the wind takes on different personas, like a god that is capable of assuming any living form. I still often think of it as the breath of a dragon, though it usually feels more like ice than fire. Its gustiness, its ability to surround and overpower me, seem to be of mythical greatness; yet it is undeniably *real.* I do not bundle up as carefully as I used to when I go out; to tell the truth, I now look forward to the cleansing power of heavy blows. But unlike the old-timers, I will never think of the ocean as the back side of the islands. It is the front line of battle, the front line against the wind.

Over the past dozen years I have been in perhaps a hundred windstorms here. A few have stayed in my thoughts. One of my clearest memories is of an August day when I stood on the back porch of my little wind-haunted house and waited for a hurricane to roar through.

It was 1986, the year of Hurricane Charley — a runt, as hurricanes go, but with gusts to eighty-five miles an hour. A friend had come over to visit my husband and me with his dog, a Chesapeake Bay retriever. The storm, passing offshore, was throwing off east wind and was not expected to do much damage. Even so, no one wanted to be out in it. It was enough to stand on the leeward side of the house and watch the myrtle bushes being shaken like rag mops.

That summer a pair of Carolina wrens had built a nest in the pump house and raised several broods. There were still chicks in the nest when the storm hit. In the excitement I had forgotten about the wrens, when I saw a quick movement under the dilapidated table where we cleaned fish.

An old beach chair was folded and propped beneath the table. Leaning over, I could see an adult wren clinging to the chair. He was soaked from rain and, judging from his hunched posture, too exhausted to move even as far as the pump house. He knew we had caught him off his home base, but he did not seem to care.

The others noticed the wren the same second I did. Nobody moved, not even the retriever, although he eyed the wren with a lazy spark of interest. Nobody did anything except look out at the wind and rain. We stood on the back porch, an unlikely alliance — two men, a woman, a dog, a bird — each of us snagged momentarily from the flow of our normal lives, refugees from the wind.

TWO

Creation's First Food

"IN THE BEGINNING God created the heavens and the earth." Deep in the collective imagination of Western culture is an image of what the world must have been like in that time, an era of murky formlessness, of penetrating cold, a black spaciousness beyond comprehension and without meaning. Nothing crawled or flinched on the whole still globe. Smells did not exist, nor did sounds. There was no thought, or hope. And then, we are told, the spirit of God moved across the waters, in the form of wind.

It is blowing nearly forty today, and I am standing on the north end of Roanoke Island — the lee side of land in this southwester — watching wind pass over water. With a thick wall of live oaks at my back I can taste the exhilaration of the storm without suffering the usual physical punishment. Limbs thrash above me but do not break. Leaves sigh between gusts, then roar as they are caught anew. The bleak gray surface of Roanoke Sound is tickled by stray breezes; small dapples near shore give way to rolling breakers a half-mile out, where the wind swoops down full force. It is the fleeting patterns close to land that most intrigue me. Swinburne wrote of watching "the wind's feet shine along the sea." To me, though, these ripples look more like trailings left by the hem of a long gown. They come in curling swaths, and the swaths come three and four abreast. There is something much larger than me aloft out there. I am glimpsing, I believe, the garments of a passing god.

THINK OF WIND as God and God as wind. Think of the fluid element in which we live, on which we depend, as a divine current that envelops us, swirls through us, and joins us to a great, organic, barely fathomable whole. To think thus is to become part of a tradition that reaches back to the earliest stirrings of humankind.

The Hebrew word *ruwach* can be translated to mean spirit, breath, or wind. *Strong's Exhaustive Concordance of the Bible* lists its primary definition as "wind; by resemblance *breath*, i.e., a sensible (or even violent) exhalation." Other connotations include air, blast, spirit, and whirlwind. *Ruwach* is the breath of life that God steals from humanity with the great flood of Genesis; it is the east wind that brings a plague of locusts to Egypt. Moses calls on *ruwach* — an easterly breeze — to part the Red Sea. In the Old Testament *ruwach* is used 27 times for breath, 238 times for spirit, and 106 times for wind. The animate force it portrays is often angry, often heroic, and given to both bestowing life and snatching it away. The Hebrew word *neshawmaw*, too, connotes not only "a *puff*, i.e., *wind*, angry or vital *breath*," but also "divine *inspiration, intellect*, . . . soul, spirit." It was *neshawmaw* that God breathed into Adam, bequeathing him both life and intelligence.

It is little wonder that the authors of the Old Testament perceived of wind as an agent of a nurturing yet jealous and vengeful God. In the biblical Holy Land (and throughout the world), the force and direction of the wind largely control the level of moisture in the air. As a result, the wind's mood through the seasons often dictated whether the authors of the Bible feasted in comfort or suffered the hardships of severe deprivation. In the dry months of summer, steady, moisture-bearing winds from the Mediterranean lightened the oppressive burden of heat and enabled threshers to winnow grain. Between summer and winter the region was beset periodically by the sirocco, a scorching southeasterly that strikes in violent blasts and often brings curtains of dust; it has been compared to the Santa

Anas of southern California. (The sirocco may be the wind that collapses the house of Job's eldest son, killing all ten of Job's children.) In winter the wind shifted to the west, and unstable air masses spawned thunderstorms, which, despite their frightening violence, were welcomed for their precious rain.

Frequently in the Old Testament, wind is a symbol of unruly, untamable power. God rides on the winds in the clouds and unleashes breezes to do his bidding. The wind has no visible source — Jeremiah claims it is kept in God's storehouses — and once it passes, its course cannot be tracked. It is often evil, but even then its work is holy, for Satan does his mischief under God's watchful eye. At times the biblical wind symbolizes transience and foolishness. In Ecclesiastes the prophet declares that many of humankind's most cherished tendencies, among them striving for money and success, are stupid, fruitless endeavors. He likens them to "chasing after the wind."

The Greek equivalent of *ruwach, pneuma* ("a *current* of air, i.e. *breath*"), more often signifies the spirit of God and less often describes the actual physical movement of gases. When Christ quiets the tempest at sea as his disciples watch — "even the winds and sea obey him!" — the term used is the Latin word *anemos,* which derives from a word meaning air and has a distinctly inanimate connotation. Nevertheless, wind in the New Testament is often enlisted to carry out God's will. On Pentecost "suddenly there came a sound from heaven like the rushing of a mighty wind, and it filled all the house where they were sitting." Even today the Holy Spirit is said to move through a crowd like the wind. Near the end of Acts a "tempestuous wind, called the northeaster," drives the ship carrying Paul and his captors aground on Malta, but, as an angel has foretold, every man is saved. In the vision of Judgment Day in Revelation, four angels stand "at the four corners of the earth, holding back the four winds of the earth," and sparing the world from destruction.

FOR NEARLY three thousand years Judaic culture and its Christian progeny have steadfastly separated God from "nature" — that is, the forces and creatures that man and woman evolved from. This seems especially curious when one considers the inherent characteristics of the invisible, formless, capricious wind. It is capable of tender caresses yet is strong enough to destroy any being or structure on earth. It has the power to arrive unbidden, to slip through the cracks in our houses, in our lives, and hurl us to the ground. All this makes it more like the god we purport to worship (every one of us, in our myriad religions) than anything else in the human realm of consciousness.

Shortly before the birth of our son I went one afternoon to a favorite beach near our home. It was late spring, and hot. My sore and swollen ankles throbbed; I could not take the walk I had planned. The pressure of the baby's descending head sent bolts of pain through my groin. Worse, I was burdened with personal problems and doubts. I had recently failed at a writing project; weeks of work would have to be thrown out. A friend had badly betrayed me. I did not have the energy to think, much less to chart a course through my grief.

The ocean did not seem capable of providing its usual solace. Gray breakers slapped halfheartedly against the shore, knocked nearly flat by a southwest breeze. Flies buzzed at my neck. Standing in the wash with aching legs, a burning wind at my back, I wanted to walk into that ugly, languid surf and submerge myself forever. I wanted to lie down in the sand but doubted my ability to get up again. So I merely stood with my eyes closed, brooding, miserable, foolish.

A current of cool ocean air woke me. With a simple shift of wind to the east, the world changed. In minutes the waves shaped themselves into pearly flutes. The sun on my arms felt as soft and welcome as flannel. The flies vanished. Your child is going to be wonderful, something said, and it will be born in a matter of days.

We go sailing now, my family and I, as often as we can. For me it is a form of renewal and prayer. Our boat, filled with the personality granted it by wind, reminds me of a great horse. The breezes joust with us. They test my mettle as a sailor. In Roanoke Sound one afternoon a northwest wind builds without warning, until the pleasant, rocking waves brim with whitecaps. The boat is solid but old; I'm wary of sailing it in winds stronger than fifteen miles an hour. It pounds through the gray, frothing swells, responding less and less to the tiller. I am no longer in charge. In the back of my mind is a story about a local doctor whose sailboat capsized in these waters, in winds just like this. His wife drowned. I try to come about, but the boat will not turn. We are being carried due east, into rougher waters and heavier breeze. I take a deep breath, haul the sail in close to build up speed, and veer hard into the wind. A wave slams the hull and breaks over the deck. The sails flop, empty. The boat teeters on the edge of the wind for a heart-stopping moment and swings around.

Relieved, I retreat to the more protected waters of Shallowbag Bay. Now the wind promises me an easy passage back to shore, but it glances off a point of land and forces me to fall off my bearing. The boat keels wildly; I bring it about for another tack. The damp rope of the mainsheet burns into my hand. Try again, the wind says, I dare you. I take the challenge and win.

I pull up to the dock with the sense of having wrestled an angel.

ON OCTOBER 29, 1991, my husband and I woke in the minutes before dawn to a vigorous wind. The trees whooshed; pine twigs fell on the roof like shrapnel. We looked at each other, eyes wide, and flopped back into bed with the covers pulled completely over us. It was hardly the most intense wind we had ever experienced; the gusts reached only into the upper twenties. Still, there was an ominous charge to the atmosphere, as if something much bigger were brewing.

That afternoon the surf was the roughest I have ever seen it. Oddly, a hurricane passing offshore had combined with a local northeaster to push the tide ten feet above normal. The water was not blue, but white with foam. Massive breakers extended to the horizon. As Jeff and I stood on a walkway above the beach, the height of the waves took my breath away. There was no soothing rhythm to the surf, no lull between the mountainous breakers. "It's like looking out across the Rockies," Jeff said in awe. The ocean we loved had become a country of seething, collapsing peaks.

Late the following afternoon, a full day after the wind had quit, we went back to the beach. The surf had calmed little; it still entranced me with its might. Jeff scanned the horizon with binoculars. "My God," he said suddenly, "there's a sailboat out there."

Through foamy air I could barely see the boat, a two-masted schooner under sail, lurching from wave to wave. We stood for a time without speaking. "Do you think they need help?" I asked. Jeff did not answer. Although we called the Coast Guard, we never learned the boat's fate.

FAR TO the southeast, a smaller sailboat was adrift in even rougher seas. On October 21 Janet Shaughnessy had left Norfolk, Virginia, on the *Saorsa,* a thirty-three-foot, single-masted craft built and skippered by her friend George Ross. The pair were bound for Bermuda, which they hoped to reach in six to eight days. Shaughnessy is a slight, soft-spoken artist for a newspaper, and by her own description not much of an adventurer. She had never been sailing.

Even before the trip, Shaughnessy says, strong winds could vastly affect her moods. "I've never been very fond of wind," she says. "It makes me feel kind of," she pauses, searching for a term, "messy. Vulnerable and stripped down. I always avoid it, so on this trip to be completely surrounded by it required a sort of surrendering."

For more than a week *Saorsa* labored to make progress through squalls and thunderstorms that pushed her far southwest of her destination. With no motor, she traveled at the whim of wind. On Monday, October 28, Shaughnessy awoke at 4 A.M. to beautiful, calm weather. "I could see the moon and the sun in the same sky. We thought, We're finally going to get there." But toward evening the wind picked up and the seas built. The couple tuned their radio to a Bermuda station and learned that the islands were bracing for the close passage of Hurricane Grace, a category-one storm with winds to eighty-five miles an hour. "Everything stopped for us," she said.

By then they had sailed nearly into Grace's northern edge. The ocean was too rough to risk going forward to lower the sails, which were double-reefed. As darkness fell, they closed themselves into the cabin and put on floatable survival jackets. The boat pitched and rolled, throwing them against the walls. The wind was deafening. "There was a kind of cycling," Shaughnessy says, "where we'd hear something like the roar of an engine — the wind in the sails, maybe — and the boat would pick up speed and tip over. Then it would right itself and the roar would start gearing up again." Twice the mast went completely under. Water rushed into the cabin. "I thought, This is how people die."

They lay in the cabin, listening, talking little, holding hands. They were surprised at their calm. Shaughnessy decided silently that if need be she could surrender her life without regret. Ross suggested a plan for getting into the life raft if *Saorsa* began to break up. He would go first to inflate the raft, and Shaughnessy would wait in the companionway with bottles of fresh water. "Neither one of us wanted to do it," Shaughnessy says. "The prospect of jumping off *Saorsa* into that ocean was terrifying."

By dawn Grace had passed. The wind speed dropped to about forty-five, but massive waves assailed them from every direction. All they could see of the sky was a small gray swatch directly overhead. In the light of day Shaughnessy felt more hopeful, even though the hurricane had shredded the mainsail. She made Ross a peanut butter sandwich, "which was no small feat, the way we were still rolling around." Ross, assessing the height of the waves from the deck, called down, "This is manageable." Nonetheless, with no mainsail they could do nothing but drift and try to keep the boat upright. Ross activated an emergency beacon. "I think it was hard for him, admitting defeat like that," Shaughnessy says. A few hours later a Coast Guard rescue plane located them, but violent, colliding waves prevented the rescuers from taking them off the boat.

They worked shifts on the tiller, struggling to prevent the boat from capsizing, delirious from exhaustion. Crews from rescue planes talked to them by radio, assuring them that they would soon be safe. A passing tanker stayed close to them that night, and its pale green running lights, appearing and disappearing behind the black swells, soothed Shaughnessy's nerves. In the wee hours, as she manned the tiller she felt a tall, calming presence behind her. "I swear it was a guardian angel helping keep me awake," she says. The sky cleared, and the constellations shone as brightly as distant fires.

On Wednesday Ross and Shaughnessy declined an offer by the captain of the tanker to take them off *Saorsa*. "It would have meant abandoning her, and Geordie just couldn't do that," Shaughnessy says. She was more tired than she had ever felt, and her emotions bounced from hopefulness and exhilaration at having survived to despair when once again it began to rain. A Coast Guard cutter reached them that afternoon. Along with a tow line, the crew sent over two steak dinners. "I've never tasted food like that," Shaughnessy says. They reached Bermuda the next day.

Three years later I came across an old newspaper story about Shaughnessy's adventure. Still haunted by the sight of the unknown schooner off the Outer Banks, I bent to a whim and called her. What was it like, I asked, to be hundreds of miles from home, adrift in that wind, those seas?

Shaughnessy was quiet for a moment. "I don't think I was right for about a year afterward," she said. But while the wind was all-powerful, while she and Ross sometimes addressed it as a presence, she never thought of it as a god or even as an agent of God. It was she and Ross and *Saorsa* versus Hurricane Grace.

I have heard this sentiment echoed a half-dozen times in conversations with people who have survived windstorms of various sorts. Wind is not regarded in Western culture as a divine force — although fishermen, caught offshore in a lashing northeaster, will call out half-jokingly, "Okay, God, we know we're sinners. You can let up any time now." One Hatteras Island captain, notorious as a blasphemer, has been known to challenge God during heavy blows. "I saw him out on the foredeck, waving his fist and hollering," a crew member told me. "Wicked winds, and the waves higher than a two-story house. When he came back to the wheelhouse I asked him what he'd said. 'I told the Lord, Goddamn it, you've sunk boats out from under me twice before, but you can't have this one. You're not going to beat me this time.' I didn't fish with him after that." When pressed, however, no one I spoke with would ascribe to wind the full potency of God. It is a test sent by God or a random disaster that their faith in God safely brings them through. It is all-powerful, yes. It sometimes seems alive. But it is not God incarnate.

In this belief our Judeo-Christian account of the world differs sharply from aboriginal religions, which are heavily grounded in nature. In most mythologies the winds are gods and goddesses who journey earthward and often meddle in the affairs of humans. Perhaps we are simply frightened by the thought of confronting Yahweh, of being brushed every day by a divine physical presence. Perhaps we are too cowed to consider that God is not merely seated in a remote heaven, but whirling in the air around us.

A tornado spins through a Hatteras Island marina, destroying five commercial boats, robbing their owners of a half-year of pay. Another forms in Croatan Sound, leapfrogs over Roanoke Island as startled residents watch, and goes east to devour a swath of maritime forest in Nags Head. Survivors of tornadoes and hurricanes often seem dazed and disoriented, like people who have glimpsed another world. In an instant their lives have been changed, their fragile mortality driven home. They wonder: Was it a god that touched me, or a devil? Or was it simply particles of air flying at unreckonable speed?

IF WIND has not delivered God to us as a palpable being, it nevertheless brought the early Jews, and the rest of humanity, an earth of order and life.

In the sixth century B.C., just before Genesis and other early chapters of the Bible were likely written, the Israelites spent between fifty and seventy years imprisoned in Babylon. Each spring the Babylonians held a festival to mark their new year, and on the fourth day celebrants recited a creation epic known as the *Enuma elish* (the title is taken from the tale's first words, "When on high"). The epic tells of a great struggle between the parent gods, Apsu and Tiamat, and their unruly sons, who "troubled the mood of Tiamat / By their hilarity in the Abode of Heaven." Apsu and Tiamat are of monstrous size but lack the power to move about; they represent the inert forces of subterranean fresh water and salt water. Over time they give life to a sky god, a god of flowing water, a god of wisdom and spells, and several others. But once these children exercise their freedom of movement, Apsu and Tiamat grow deeply envious.

Apsu approaches Tiamat and bids her to help him annihilate their sons. As would any good mother, Tiamat refuses. "What? Should we destroy that which we have built?" she asks. "Their ways are indeed troublesome, but let us attend most kindly!" So Apsu devises a plan to kill his sons on his own. Unfortunately, the younger gods learn of the plot and slay Apsu before he can act.

When news of Apsu's death reaches Tiamat, she calls upon a force of scorpion men, roaring dragons, and lion demons, all "sharp of tooth and unsparing of fang," to march into battle against her rebellious offspring. The sons gather in fear, debating what to do. Ea, the wisest, takes a wife and begets Marduk, a storm god and the paramount deity of Babylonian religion. Another son creates the four winds and sends them to plague Tiamat, raising waves on her surface and churning dirt into her waters. The goddess becomes too agitated to sleep. Furious, she assembles her monsters, which are commanded by the most ferocious of them all, a god named Kingu, on whose breast Tiamat fastens a "Tablet of Destinies."

Marduk — who alone is not cowed by Tiamat — mounts a storm chariot pulled by four fleet, poison-breathing horses and attended by winds and hurricanes, lightning and thunder. He approaches Tiamat with a great net to ensnare her. The enraged goddess opens her mouth to consume him. But Marduk has anticipated this, and he sends into her a fierce wind that distends her body and makes it impossible for her to close her jaws. He shoots a poisoned arrow into her, which pierces her heart. After she dies, Marduk "split[s] her like a shell-fish into two parts." From these he creates the sky and the earth. The Tablet of Destinies he claims as his own.

It is a familiar tale, one of a god who separates "the waters which were under the firmament from the waters which were above the firmament." Imagine the Israelites: exiled, resentful, questioned by their children about their own God, and forced to listen to recitations of the *Enuma elish* year after year. The creation story that arose in Jewish culture bears a striking resemblance to the Babylonian epic in its use of a wind that imposes order over chaotic waters.

Creation myths in which the wind plays key roles occur throughout the world. The Hindu creator-god sent a primordial wind to sweep the waters of earth during the genesis; so did the god of the Quiche Indians of Mexico. In Japanese mythology, breezes performed the final task of creation by dispersing the mists that enshrouded the islands. The Indic Rig Veda describes the wind as the breath of Varuna. (In the meditation practices of many Eastern religions, the awareness of one's breath is of supreme importance, for breath joins each person to the whole of creation.) The ancient Egyptians, the North American Eskimos, and the Aztecs all had words that at once signified breath, wind, soul, and a great spirit. Of course, tales of virgin births, worldwide deluges, and the theft of fire from the gods also developed among diverse and far-flung tribes. Joseph Campbell celebrates the intricacies of such stories, which he describes as "appearing everywhere in new combinations while remaining, like the elements of a kaleidoscope, only a few and always the same."

In his book *Heaven's Breath,* Lyall Watson describes the winds as "the direct offspring of Mother Earth. They are, however, no ordinary children, but wayward, somewhat rowdy progeny over which there is little or no possibility of control." Watson regards the wind as a giver of seed that enables the female earth deity to create other beings. He notes that ancient peoples attributed to the wind powers of insemination. Egyptian vultures were thought to be entirely female; the wind fertilized them. Homer wrote of mares who stood with their backs to the north wind and conceived foals so light of foot they could gallop across fields of standing grain without dislodging a single seed. The great Iroquois warrior Hiawatha came to life in his mother's womb after a visit from the holy west wind. Watson finishes by describing wind as a "potent and creative force, a very seminal thing, but lacking in social graces. The winds, in other words, are unquestionably male."

I listen to a breeze whooshing through the pines over my house and imagine a great rustling of taffeta, a passing over of giant petticoats. Why shouldn't the wind, with all its personalities, have both male and female faces? The Greek goddess Athena was the deity of fresh air and gentle zephyrs. A breeze that keens through a seaside cavern may sound like a siren, one of those singing nymphs who lured Greek and Roman sailors to their deaths. Unquestionably male? I shake my head at Watson's words. No one can hang a gender on such a changeling, any more than man or woman can assign a sex to God.

IN THEIR explanations of why the wind blows, ancient peoples took into account the diverse nature of breezes, and they drew heavily on religious beliefs that attributed divine powers to many features of the natural world, especially to mountain peaks.

The tribes of central Asia believed that the wind originated from a hole in a mountain, the exact location of which was rather vague. In the thirteenth century a Catholic friar exploring the Dzungarian Basin of northwest China found "a mountain in which is reported to be a hole from whence, in winter time, such vehement tempests of wind do issue that travellers can scarcely, and with great danger, pass by the same way." Later explorers found an extinct volcano rising from Lake Ala Kul and, nearby, a pass in the Ala Tau Mountains known as the Dzungarian Gate. Winds flowing from the lake are funneled through the pass, which in a short distance narrows from a width of twenty-five miles to six. The Mongol and Turkic nomads of the region still teach that the volcano generates furious gales and dispatches them through the gap in the mountain walls.

In the remote mountainous rain forests of North Borneo, the Tempasuk Dusun people pay homage to a god who works as a blacksmith, forging and reforging souls. When souls wear out, as all inevitably do, the blacksmith transforms them into winds and places them in a hole in a hill. The New Zealand god Maui is capable of both riding the winds and imprisoning them in caves, corking their entrances with large rocks. Despite Maui's power, wrote the anthropologist Edward Tylor in 1871, "he cannot catch the West wind, nor find its cave to roll a stone against the mouth, and therefore it prevails."

Stories with the same crucial elements sprang up independently on this continent, too. The Iroquois of the eastern woodlands believed in a spirit named Gäoh that kept breezes in a Home of the Winds deep in the mountains. In his 1899 book, *The Eskimo about Bering Strait,* Edward William Nelson relates a myth in which a childless couple carve a doll from a special tree. The doll comes to life and travels to the sky wall, where he finds a series of holes covered with gut-skin cloth. He opens each hole for a few minutes — just long enough for a fierce wind to enter the world, carrying with it such blessings as reindeer, trees, bushes, ocean, rain, and snow. Then he replaces the covers and admonishes the different winds to "sometimes blow hard, sometimes light, and sometimes do not blow at all."

In the myths of the Abenaki tribe that peopled the coniferous woodlands of what is now northern New England, a trickster named Gluscabi sets out to stop a strong wind that is making it difficult for him to hunt ducks. He walks into the wind for days until he comes to a mountain peak where an eagle perches. This great bird moves the air all over the world by flapping its wings. As Gluscabi approaches the peak, the gales grow so strong that they blow off not only his clothes but all his hair. Gluscabi manages to deceive the eagle, bind it with bark rope, and throw it into a crevice. "Now," he says, "it is time to hunt some ducks." But the world to which he returns is hot and uncomfortable, and the bay where he wishes to hunt has grown stagnant. Chastened, Gluscabi goes back to the crevice, retrieves the eagle, and frees it.

Most Native American peoples envisioned the wind as four distinct beings, a belief they shared with the ancient Greeks. The Greek god Aeolus is given domain over the winds by Zeus, who originally housed them in a cliff in the Tyrrhenian Sea but found them too rambunctious to be easily controlled. Aeolus keeps his charges — Zephyros, the gentle west wind; Boreas, the chill north wind; Notos, the southern rain-bringer; and Eurus, the ill-tempered east breeze — locked in a cavern until he is visited by Odysseus. Charmed, Aeolus dispatches Zephyros to carry Odysseus safely home, and he gives the adventurer the three stormier winds tied up in a goatskin sack. The mouth of the bag is so tightly closed that not a single puff of air can escape. Unfortunately, Odysseus falls asleep within sight of Ithaca, and his greedy companions open the bag, hoping to find gold within. The ensuing tempest tosses the ship mercilessly for many days, carrying it back to Aeolus (who refuses to help the beleaguered sailors) and finally to the land of the Laestrygons, a tribe of cannabalistic giants.

I read these myths eagerly, pulled by common threads from source to source. I am deep inside Campbell's kaleidoscope. I know wind to be nothing more than a mix of gases pulled by suction around bumps and pits in the atmosphere, yet I feel as if I am part of a tradition that predates recorded time. And among my contemporaries I am in good company. I have two friends, Barbara and David, an atheist and an agnostic, whose passion for windsurfing approaches religious zeal. They joke about making sacrifices to the wind so it will blow, not just ten or twenty miles an hour but thirty or forty. "When I'm in the ocean and everything clicks," David says, "and I'm hauling along with only a few inches of my board in the water, I just start screaming 'Yaaaa!' I feel like I'm part of something much bigger than me or anything I know. And I'll pass someone else, and they'll be screaming too." Is that a kind of prayer? Dave laughs and shrugs. "I wouldn't call it that, but it's the closest I'll ever get to prayer."

I read and read. The Chinese god of wind, I learn, is an old man called Feng Po who sports a long white beard and a blue-and-red cap and carries the breezes in a sack slung across his back. When he wants the wind to blow from a certain compass bearing, he points the mouth of the sack in that direction. This deity is also said to take the form of a dragon (a dragon!) called Fei Lien. I like the notion of God in the form of a powerful, moody beast that sends forth torrents of air from the edge of a world I perceive as only ocean.

Many aboriginal cultures believe wind to be the restless souls of the dead — in the case of storms, the warring or unhappy dead. Among the Sea Dyak of Borneo, the Wind Spirit serves as a messenger to the underworld whenever a person dies. He begins the journey by traversing an endless plain and climbing a tree to search for the right way. He must choose his route with great care, for there are seventy-seven times seven roads to the realm of the dead. Once he has discovered the best road, the Wind Spirit rushes down it in the form of a hurricane. The dead quake with fear at his approach. "Someone has died," the Wind Spirit says, "and you must hasten to fetch his soul." The dead joyfully set out in a boat and row to the person's house. They seize the soul, which cries out once, then goes with the spirits in peace.

DEEP IN the red-rock, breeze-chiseled country of the American Southwest live a people whose religion may be more steeped in wind than any other in the world. The Diné, or Navajo, believe humans are brought to life by *nilch'i,* the Holy Wind, which leaves whorls on the tips of our fingers and toes, whispers words of advice in our ears, and dictates the number of days we spend on earth.

For his 1981 book *Holy Wind in Navajo Philosophy,* James Kale McNeley interviewed ten Diné about the various powers attributed to the winds. Eight of the men interviewed were elderly singers who performed the ceremonials and healing chants central to the Navajo religion, including Windway, the rite used to cure ills caused by winds, snakes, cactuses, clouds, the sun, and the moon. This gives their words special weight, since singers are charged with preserving tribal tradition.

Unlike the Western concept of wind, the Diné wind is a spiritual force, one of the holy beings. The Diné creation stories begin in an underworld with the "misting up" of lights of different hues from the horizons — in the east, a light as white as the edge of dawn; one blue like the sky at midday in the south; in the west the yellow glow of twilight; and a deep black beacon in the north. Then came the Wind. "The Wind has given men and creatures strength ever since," a Navajo man told Father Berard Haile in 1933, "for at the beginning they were shrunken and flabby until it inflated them, and the Wind was creation's first food."

Not only was Wind the source of life and breath, but it bestowed on humans (as in the biblical Genesis) the power of thought. At first all the tribes wandered aimlessly through the underworld, incapable of making plans, until they encountered Wind in the form of a human. "I will see for you," Wind told First Man, First Woman, Talking God, and Calling God. "I know about what is in this Earth and what is on it. I am Wind!" The Peoples could not talk, but Wind taught them how. He positioned himself in the folds of their ears so he could always be close to them to act as conscience and guide. And the Peoples found that when Wind told them something would come to pass, his prediction invariably came true.

In time several Holy People climbed up to the surface of the Earth through twelve reeds. Wind emerged with them in the form of four cardinal breezes, two male and two female. These came to serve as the breath of the sacred mountains that rim the horizon of the Navajo cosmos. The most holy of these, a female wind, went to live in the east. At the same time, two winds were born of the earth, two of water, and two of clouds. These met and mingled, and "six winds [were] formed above and six below," McNeley writes. "We live between these, . . . all of which affect us and some of which cause difficulties and sickness. . . . There is only one Wind, [but] it has twelve names."

After the emergence of the Holy People from the underworld, Earth Surface People were made from primary ingredients such as soil, lightning, and water or, according to another version, from corn mixed with jewels like turquoise and white shell. These elements gave the Earth Surface People substance, but a holy breeze known as Little Wind or Wind's Child gives them breath and helps them stand erect. A person's posture, balance, and ability to speak all are gifts of the winds dwelling within him. "It is only by means of Wind that we talk. It exists at the tip of our tongues," a singer told McNeley. Winds also "stick out" from the soft spots on the tops of our heads and the curving lines on our fingers and toes. "These whorls at the tips of our toes hold us to the Earth. Those at our fingertips hold us to the Sky. Because of these, we do not fall when we move about."

When a child is conceived, it takes a wind from its mother and another from its father. These lie on top of each other — as the winds of early creation are said to have done — and merge to become the child's own. Four months after conception this "Wind that stands within one" causes the child's first movements. At birth, with its first breath, the child receives another wind, sent from a holy source, that will guide his or her life. Early missionaries to the Navajo believed that the internal wind resembled a soul, but in fact the concept is much more complex and not easily translated into the vernacular of Western culture. The inner wind draws constantly from other winds and attaches the person to the entire swirling, holy atmosphere. "That within us stands from our mouth downward, it seems," a singer said. "We breathe by it. We live by it. It moves all parts, even our hearts."

OF ALL the Navajo beliefs, none is more provocative than the notion that the winds that enter a person are responsible for his behavior. An individual's temperament turns not on upbringing or on any misfirings within his neurological circuitry but on which winds he allows to guide him as he moves through life. This ties individuals even more strongly to the natural world, for a person's beliefs and actions — indeed his very mind — belong not to him but to the holy air.

Many Navajo believe that the wind acts as a messenger to the four sacred mountains, as well as to the sun and the moon. When a person thinks evil thoughts or performs a bad act, the wind bears the news to the Holy Ones. It does this in the form of Little Winds, the conscience that also lives in the ears of individuals and whispers advice to them. In some of the Windway ceremonials, sand paintings are done of the cardinal winds, portrayed as tall, geometrically drawn people; the Little Winds are shown murmuring in their ears.

The Holy Ones send particular Little Winds to each person. These are different from the wind that squires the person through life from birth to death. (All the winds, however, are seen as small currents within the holy atmosphere.) The Little Winds admonish the person to act and speak with virtue, and they protect him from danger and harmful outside influences. If a person listens to his winds, he is said to walk in a good way and to lack faults. If he consistently ignores them, they will leave him. Without the support of the Little Winds, the wind standing within the person weakens, and he is filled with Evil Winds. "Over there it seems an evil one is waiting," a singer said. "Here, the one that he really lived by seems to stop [working]. Then the evil Wind runs in here [between the eyes]. At this point bad things happen." The person's thinking becomes flawed — he stops planning ahead, for example, or lets himself be overcome by jealousy — and he acts foolishly, recklessly, even maliciously.

A person with truly evil intentions may invoke witchcraft to try to gain control of another's wind. Or she may attempt to entice the winds to lead dangerous animals to attack an enemy. If the enemy's internal wind is stronger than her own, however, her efforts to cause harm will ultimately fail. If she does not mend her ways, the wind standing within her will weaken further. Her physical strength will wane, and she will fall ill. In the end she will die. The Navajos fear the spirits of those who die young, for a youthful death is evidence of a harmful wind within. Such winds may become evil dust devils that sweep the desert, rotating sunward, the opposite of the good winds, which turn in the same direction as the sun and the moon as they travel through the sky.

To die young in Navajo culture is to die damned. Ah, but to live to old age — that is a sign that one has followed the counsel of the Little Winds, and through them the Holy Ones. The wind standing within the person dies as it should, peacefully, in concert with body and mind.

GIVEN THE rich variety of mythologies about wind, as well as its enigmatic qualities, one might expect the earliest scientific theories about its origins to be a hybrid of religious beliefs and crude observation. But the first rustlings of cause-and-effect science marked the advent of a new religion and the inevitable death of the deities who had previously ruled the skies. While the initial scientific explanations of the wind's origins were roughly formed, in fact they were not far from the truth.

In the sixth century B.C. Anaximander, a Greek philosopher of the Milesian school, wrote that the earth, the heavens, and all within them came into being when a great primordial sea was dried by evaporation and celestial fires. All atmospheric phenomena, he continued, are the result of opposing forces — heat and cold, dryness and moisture, light and dark — brought to bear on fire, water, and air; and wind is a "current" set in motion when mists or vapors are "stirred" or "melted" by the sun.

A hundred years later, Empedocles used a simple flow tank to demonstrate that moving air can exert a force on water. He suggested that wind and moving air were one and the same. In the first or second century B.C. the Macedonian astronomer Andronikos built a water clock within an octagonal building on the edge of the marketplace of Athens. Each side of the Tower of the Winds was decorated with a frieze depicting the personality of the appropriate wind: the warmly dressed Boreas, or north wind, who holds a conch shell through which he howls; the stern, elderly Kaikia, the northeast wind, carrying handfuls of hail; the sweet-faced, fruit-bearing Apeliotes or east wind, which brings fertility and abundance to Greece, and so on. All the figures faced clockwise, the way the winds of the northern hemisphere move around the compass in fair weather. Recent archaeological evidence suggests that the tower may have been the centerpiece of a complex cosmology that linked Greco-Roman creation stories with theories of meteorology, chemistry, physics, and medicine. The Muslim world, too, may have developed a detailed cosmology based on wind. According to ancient texts, the four sides of the Kaaba in Mecca are aligned with what Arabs believed to be the four cardinal winds.

Around 340 B.C. Aristotle declared wind to be born of a vapor, a "Dry Exhalation" arising from the earth. At first blush this sounds as if he correctly intuited that winds are created by the upward drift of warm air. But Aristotle also used dry exhalations to explain the existence of comets, shooting stars, lightning, and thunder. Moreover, he insisted that wind cannot be air in motion. Instead, an invisible force must move air, for, as he writes in *Meteorologica*, "the same air persists both when it is in motion and when it is still." The sun, he believed, both pushes the winds and checks their speed. Earthquakes are caused by subterranean winds that occasionally surface through caverns, just as the rumblings in a gentleman's stomach (and, one assumes, the occasional unbidden releases of gas) are caused by "the force of the wind contained within our bodies. . . . We must think that the earth is affected as we often are after urinating — for a sort of tremor runs through the body as the wind returns inward from without in one volume."

"Some," he writes, "wishing to say a clever thing, assert that all winds are one wind, because the air that moves is in fact all of it one and the same." Such theories, he asserts, are poppycock. "This is just like thinking that all rivers are one and the same river, and the ordinary unscientific view is better than a scientific theory like this."

Despite their many errors, Aristotle's theories were strikingly original and, for the time, meticulously conceived. No one would pose a significant challenge to them for many centuries, in part because the spread of Christianity halted most scientific inquiry into the workings of the natural world. Early Christians were admonished not to question scriptural depictions of forces created by God's hand. The few new opinions on nature to be published during those dark days generally came from holy visions such as those received by Saint Hildegard of Bingen.

Hildegard was a Benedictine nun and one of the most remarkable women of the Middle Ages. Had she been a man, she might have won world renown as an intellectual. Between 1141 and 1179 she dictated numerous works to monks, who translated them from German into Latin. Many of her writings detail her visions and thoughts about God, but she also compiled encyclopedias and wrote books on natural science and the human body. The detail in these latter works prove her to have been a meticulous observer.

Describing a vision, Hildegard writes, "I looked — and behold — the east wind and the south wind, together with their sidewinds, set the firmament in motion with powerful gusts, causing the firmament to rotate around the earth from east to west." She believed that the atmosphere comprises four concentric layers, each containing a cardinal wind. East wind lies below west wind, which is succeeded first by the north breeze and, closest to the sun, the hot southerlies. The layers also hold the heavenly bodies: clouds, moon and stars, lightning, and sun all dwell in separate stories, like diverse families residing in a four-decker house. The winds leave these abodes to dictate the seasons, to bring rain or withhold it, and to do, when He wishes it, the bidding of God.

Although the church gradually relaxed its prohibition against scientific inquiry, Aristotle's views on wind held sway through much of the sixteenth century. Nonetheless, some thinkers gave the question of the wind's birthright new and engaging, if farfetched, consideration. Francis Bacon believed wind to be generated by vapors that expand their volume a hundredfold and so become air. He wrote that "the places where there are great stores of vapours [are] the native Countrie of the Windes."

Around the middle of the seventeenth century, Galileo's student Evangelista Torricelli invented the barometer and provided the world with the first instrument for measuring meteorological events. Torricelli, realizing that air flows from regions of high pressure to centers of low pressure, laid the foundation for our modern knowledge of wind. His work led to the establishment of an alliance in which observers throughout Europe compiled temperature and barometric pressure readings and compared them.

In 1686 Edmund Halley, who first tracked the orbit of the comet that bears his name, presented a groundbreaking paper to Britain's Royal Society. Halley suggested that differences in atmospheric pressure over land and over water stir the air and nudge it into motion. The trade winds on which sailors so depend, he added, are caused by the intense heat of the sun at the equator.

Halley's theories were not widely accepted for decades, in part because of conflicting views set forth by at least one other member of the Royal Society. Two years earlier a Dr. Martin Lister had presented a paper arguing that wind is caused by the breathing of the great mats of knotty yellow weed in the Sargasso Sea, the salty center of the North Atlantic gyre. Lister noted that so much plant matter collected in one place must release a tremendous amount of respiration. How else might that massive breath manifest itself but in wind?

IRONICALLY, the same era that spawned the scientific study of wind also saw a proliferation of superstitions about how it might be conjured and controlled. The most famous tales involved the sorcerers of Lapland, Finland, and Scotland, who by virtue of their geographic location were thought to be closest to (and therefore most knowledgeable about) the source of northern gales.

Tales of wind wizardry in Lapland and Finland date to 1179, and an unpublished thirteenth-century treatise by an anonymous author refers to the practice of selling winds. Becalmed sailors would purchase an enchanted thread or leather thong containing three tight knots. The first, if loosened, would bring a gentle, pleasant wind; the second a brisk, freshening breeze; and the third a raging tempest. Scottish witches tied a moistened enchanted rag around a piece of wood, knocked it against a stone, and said three times over:

> I knock this rag upon the stane
> To raise the wind in the devil's name.
> It shall not lie [cease] till I please again.

A similar chant was used to make the wind stop blowing. No one could practice the black magic of the breezes without first being baptized and then renouncing Christ. For this reason, the vending and purchase of winds were heartily discouraged by the church.

Belief in such spell-casting was bolstered both by folk tales and by the fickle nature of maritime winds. In 1560 King Erik of Sweden set out to woo England's Queen Elizabeth, but he abandoned his plans after squalls beset his fleet off the coast of Norway. His misfortune, legend holds, was caused by the mischief of witches on shore. In Shakespeare's *Macbeth,* written in 1606, the three witches give each other winds as a token of friendship. The great Gustavus Adolphus, the king of Sweden who invaded central Germany during the Thirty Years War, was said to have been aided by wind magic practiced by the Lapps and Finns in his armies.

In the mid-seventeenth century a number of writers attempted to dispel beliefs in wind wizardry. Thomas Ady, an Englishman, published a book in 1656 in which he claimed that knot-tying wind vendors were in fact "Impostors" in cahoots with "some skillful Astrologian . . . [who] can give a neer guess by the Stars, when such a Wind will arise." Nonetheless, the superstitions persisted, no doubt shored up by the firsthand accounts of such learned men as Pierre Martin de la Martinière, a French surgeon who in 1653 participated in an expedition to Lapland. In the flat, black waters near the Arctic Circle his ship was becalmed. A party of men rowed ashore to a small village, where a local wizard offered to affix a foot-long piece of cloth to the foresail in exchange for a pound of tobacco and ten silver crowns. The captain accepted the deal, untied the first knot, and the ship sailed off under "a West South-west wind, the pleasantest in the World."

Days later, de la Martinière wrote, when the wind from the first two knots had been used up, the captain untied the third. Immediately a furious northwest wind began to blow. "It seemed to us as if the whole Heavens were falling down upon our heads," de la Martinière wrote. After three days in violent, pitching seas, the ship was dashed against rocks. In desperation the crew began to pray, and the wind and waters calmed.

The art of wind conjuring followed settlers to North America, where it was absorbed into the folklore of New England and Canada's Maritime Provinces. The sailors of Maine believed that wind could be purchased by throwing money overboard. In his book *Buying the Wind*, Richard Dorson relates an 1890 tale of a Downeast captain who curses God when his ship, filled to the gunwales with spoiling fish, hits a windless patch of sea. A pious cook, disturbed by the captain's swearing, urges him to "call on the Lord, make him a love offering to show your good intent, and ask Him to send you a little wind."

The captain shrugs, lifts his eyes skyward, and says, "Lord, if you're as good as this guy says you are, ship me a little wind." And tosses a half-dollar overboard.

As soon as the coin hits the water, a hurricane blows up and sends the ship careering into the shoreline at eighty miles an hour. The captain is left alive but is buried under six feet of fish and drowned sailors. As he struggles free, the cook — unharmed and ecstatic — greets him with, "You got your wind, Skipper, you got your wind!"

The captain shifts a wad of tobacco to his other cheek, surveys the damage — "the bodies of his men laying in the surf, the fish in the trees, the riggin' up and down the shoreline" — and nods. "Yes," he says, "but by God if I'd known His wind was so damn cheap, I wouldn't a' ordered so much."

ONE DAY after I stood watching wind play across the water on the north shore of Roanoke Island, all that remains of the gale is a wrack line of dried reeds tossed high on the sound-side beach and a lumpy blanket of pine cones in my yard. The air resonates with freshness. Usually I feel washed clean and newly filled with life after a storm, but I am tired from a night of windy dreams. I sit on the back porch and watch a flock of cedar waxwings dart through the pines. They move ceaselessly, little wedges of darkness and light, flitting from deep shade to full sun. Their soft, trilling calls mingle with the soughing of leaves.

I rock on my haunches, a mug of coffee cradled in my hands, and think back over stories of wind. Two in particular refuse to be shaken from my mind. The first is a fable told by a Palestinian on the West Bank about a day of judgment in which God will send forth a yellow wind from Hell. The Arabs call it *rih asfar.* A scorching east wind, it comes once every few generations to set the world on fire. People will hide from it in the coolness of caves, but it will seek them tirelessly, especially those who have been cruel to others. Licking with hot tongues through cracks and crevices, it will roast the unjust in their lairs. Afterward the mountains will crumble into a powder that will cover the land, the Palestinian said, like yellow cotton.

The second, a kinder story, comes from the Indians of Mexico and Central America. It tells of a period just after creation when people have plenty to eat but no real reason to live. Love exists only in the form of a beautiful maiden, Mayahuel, who is watched over by an ancient goddess. One day the wind god Ehecatl finds Mayahuel and her guardian sleeping. He bestirs the maiden with his breezes and convinces her to accompany him to earth.

I think of love brought earthward by a wave of moving air, caressing the multitude of humanity, seeping unbidden into our pores. It occurs to me that love is one of the few things as ubiquitous as wind. I remember a maiden who lived in Galilee two thousand years ago; I wonder if she shared my fondness for praying on unsheltered hillsides during hard blows. Might she have gone into the wilderness one windy night, seeking holiness, hoping to glimpse the face of God? Alone but for the howling gusts, might she have conceived a child who would bring a message of love to all the world?

THREE

Guiding the Hand of History

IT IS JUST before daybreak, and we are marooned on an island in the barrier system south of Cape Hatteras. That is, I think we're marooned. The light is still too weak for me to gauge the extent of our plight. Our family came to Portsmouth Island with a group of friends two days ago and planned to take a ferry home this morning. But just after midnight the shore was ambushed by a northeast wind, one of those ruffian, out-of-nowhere spring blows. If it gets stronger with sunrise, we will not be leaving.

My husband and son sleep despite the surf that pounds the beach fifty yards from our beds, despite the rattling of two north windows in ill-fitted, half-rotted frames. Our rented "rustic cabin" is little more than a fishing camp fashioned thirty years ago from tarpaper and scrounged wood. It was comfortable enough in yesterday's light southwesterly, but I do not trust it to hold up in a major storm.

I pull on a sweater and open the front door just enough to slip through. The force of the wind makes me draw in my breath. I sprint down the steps, round the lee side of the cabin, and jog across a small field to the neighboring shack. A light is on. My friends Marcia and Nancy are up making coffee.

Marcia's eyes are wide as she opens the door. "How bad is it out there?"

"I'd guess maybe thirty." We are accustomed to tossing out wind speeds like children's names. A given number, a given personality and set of problems. As if in response, the wind swirls and hoots down an old wood-stove vent.

Marcia closes the door behind her, leans against it, and sighs. Her son Patrick, bivouacked in an upper bunk, groans and pulls a pillow over his head. "Hey," I tell him, "you may get to miss school tomorrow. You may get to miss school for a week."

Nancy laughs. "Let's not be dramatic," she says. I stick out my tongue at her. I am enjoying myself, though we hardly have enough food to last an extra day.

"I wonder how often people do get stranded out here," Marcia says. "Not as much as they used to, I bet." Probably not. But I'm excited. I want to be held captive by the wind, as long as it does not involve too much discomfort or danger. And for an hour after the sun spills over the horizon, it seems that I may get my wish. The ocean, raked shoreward, tosses wave after frothy wave onto the beach. But the marina where our cars are parked is on the mainland to the southwest of us, which means the thirty-foot boat that serves as a ferry can run with the swells of Core Sound at its stern. At 10:30 the wind slackens a bit, so we toss our gear onto the deck of the ferry and begin the wet ride toward home.

But what if the wind had not let up? What if it had blown as a banshee howls, a bearer of ill news, for days without cease? What if the ocean had spilled over the dunes and covered the islands, as it has in the past?

We like to think of ourselves, in our modern houses, our engineered office buildings, our broad-beamed, seaworthy ships, as safe from the day-to-day whims of the wind. Except for extreme cases, the odd hurricane or tornado, the hundred-year storm, we like to believe that wind has as little to do with our personal and societal histories as, say, the number of shooting stars passing over our own slice of sky. Even now we would be badly mistaken to think so.

To a large degree, the climate of the world is shaped by the patterns of prevailing winds, especially by the westerlies, which flow in undulating waves across the earth's temperate latitudes, moving most of its atmosphere. Scientists believe the wanderings of these great air masses have played a pivotal role in the climatic fluctuations of the past. Occasionally an elongated loop forms in the westerlies of the upper troposphere (the most closely fitting of the earth's atmospheric cloaks), bringing them closer to the equator than what we consider normal. At such times glaciers resume their southward creep, summer becomes a fleeting season, and seas grow too stormy for passage — while a few thousand miles east of the loop, where the storm-steering westerlies swing back north, the weather may be so hot and dry as to threaten harvests.

There is more to it, of course. Global weather patterns are tremendously complex, and shifts in climate depend on a dozen factors, including the flow of ocean currents, the surface temperatures and pressures of seas, the wobble in the earth's orbit, eruptions of volcanoes, and the extent of cover by ice and snow. But the movement of air is not simply a phenomenon that excites our imaginations and tickles our skin. It is a key player in the unfolding of life in every region on earth.

THE SIMPLEST way to gain an understanding of the global wind system is to begin at the equator, where air warmed by intense solar radiation rises and moves toward the poles. Surface winds, caught in a vacuum created by the rising air, rush toward the equator from both north and south but are turned westward by the force of the earth's spin, a phenomenon known as the Coriolis effect. These so-called easterlies (because they flow from the east) are the trade winds, the steadiest, most reliable breezes on earth. They, too, warm, rise, and drift toward the poles. All this occurs in the troposphere, the dwelling place of the world's weather.

The risen air begins to sink when it is about 30 degrees north or south of the equator — the edge of the subtropics, the latitude of New Orleans and the northern rim of Africa. A portion of the air turns again toward the equator to repeat the cycle, but some moves toward the middle latitudes. As this air travels, the Coriolis effect deflects it slightly clockwise in the northern hemisphere and slightly counterclockwise in the southern hemisphere. Air flowing away from the equator becomes the surface westerlies of the middle latitudes. As it meets cool air drifting toward the equator from the poles, it mixes and swirls to form the moving fronts that define the volatile weather of the temperate zones. In the upper troposphere the mixing air begets atmospheric waves that drift around the globe from the west.

Encircling the poles are lighter, more variable surface winds that tend to flow from the east. (This is not to say that the polar regions are zones of gentle breeze. In Antarctica, flow patterns cause air to plummet down mountainsides, reaching some of the fastest speeds on the earth.) Cold air piles up in the farthest reaches of the globe, creating strong zones of high pressure. Ultimately this cold air flows away from the poles. When it nears 60 degrees (the latitude that runs through the southern tips of Alaska and Greenland) it warms slightly and rises, spawning semipermanent systems of low pressure. At the same time, a zone of high pressure forms in the horse latitudes (including the arid zones of the American Southwest and the Middle East), where cool air sinks over land. The highs and lows — both the standing cells and the traveling, transitory systems that bring us periods of sun and storm — greatly affect atmospheric circulation by serving as peaks and valleys. Particles of air, seeking the path of least resistance, glide around the rims of the highs and fall into the sucking funnels of the lows. And pressure systems affect the direction of wind: in the northern hemisphere, air is pulled clockwise around high-pressure knots and counterclockwise into low-pressure drains; in the southern hemisphere this rule is reversed. Especially in temperate regions, air weaves nimbly from pressure system to pressure system, doing a do-si-do until it drifts back to the equator to begin the cycle anew.

This is a generalized description of surface winds; the actual flow depends on an array of factors, including the season of the year and whether the air is passing over land or open ocean. Air flows more smoothly over water, in part because a liquid surface is usually flatter than one dotted with trees, buildings, hills, and mountains, and in part because water cools and heats more gradually than land. Even the reliable trade winds deviate from their typical flow where they pass over continents. The movement of air at the earth's surface is also greatly influenced by the jet streams, the powerful westerly rivers of the upper troposphere that steer storms, and the smooth-flowing stratospheric winds, which can saw the tops off hurricanes before they fully organize and which may have a greater effect on climatic swings than scientists know, even now.

IN PREHISTORIC times, when the earth was a newly formed furnace devoid of life, it was swaddled in a soup of hydrogen, methane, ammonia, and water vapor. Temperatures were extreme, so high winds and violent thunderstorms were probably common. Beyond that, it is assumed that the regime of global winds has followed the same basic circulation pattern that exists today. Regional winds were much different, of course, because the continents were fused into one great mass. But scientists believe that during the earth's entire existence the trade winds have blown with uncanny reliance, the temperate latitudes have been beset by weather-bearing westerlies, and the polar latitudes have enjoyed light easterlies. Scientists have used studies of wind-carried sediments, in fact, to gauge the position of the North American continent during various geologic epochs. Patterns of ash-fall from volcanoes in the Appalachians and some western ranges indicate that much of the continent would have been in the belt of the trade winds throughout Paleozoic time, 600 million to 230 million years ago. But by the Mesozoic, the era of dinosaurs and flowering plants that lasted until 65 million years ago, the continent had drifted north to the latitude of the westerlies.

Throughout the earth's history, as in modern times, the presence or absence of wind would have had far-reaching consequences, in terms of both temperature fluctuations and — perhaps most important — the distribution of moisture.

Between 40 and 50 million years ago, South America and Australia broke away from the supercontinent Gondwana, leaving Antarctica marooned at the South Pole, its cupped hand extending toward the departing specks of Tierra del Fuego. Before, Gondwana had acted as a brake against the strong westerlies that flowed between 35 and 50 degrees latitude. But with the continent in pieces, the winds were able to scream with a vengeance across open ocean. (This band is still known as the roaring forties, famous among sailors for its treacherous gales and seas.)

The increased winds spawned cyclonic storms that sent a skirt of water flowing north, forming the Antarctic Convergence, an oceanic boundary of frigid water surrounding the continent. Then, during the ice ages of the past 1.8 million years, bottom and surface water from the Antarctic funneled into the sea currents that flowed north, reaching far up the western shores of Australia, Africa, and South America. The heightened temperature gradients caused by the colder currents probably strengthened the prevailing westerlies and so lengthened the rain shadows cast by mountain ranges. Patterns of rainfall were drastically altered across entire continents. Once-lush lands turned to desert. The great Amazon forests splintered into a few isolated islands, surrounded by savanna; the Congo River Valley was severed in places by north-moving fingers of the Kalahari Desert. During a number of ice ages — perhaps as few as four or as many as ten — tropical species that had lived throughout extensive forests found themselves confined to dwindling pockets, or islands, of habitat. The humid conditions necessary for the survival of these species existed only in very low areas away from mountain rain shadows. Each isolation lasted for thousands of years, and each time the forest species changed subtly, evolving slight differences from their sister species in other regions. Or so many biologists believe. The theory is controversial, and there is no certain way to test it. But the broad shifts in atmospheric circulation and rainfall may have been one of the primary causes of the great diversity of species found in the tropics during modern times.

These shifts may also have determined which civilizations prospered and which failed, and when. Since the 1950s a number of climatologists have pieced together a history of global atmospheric circulation from archaeological evidence and pollen records.

Changes in the massive drift of air from the west are a primary force in climatic fluctuation, steering the course of history with a much surer hand than any religion or empire.

The girdling westerlies reach as high as 50,000 feet and migrate over a wide area of both the northern and the southern hemisphere. At times they dip close to the equator; at other times they stretch poleward to more than 75 degrees. They contain the surface westerlies of the temperate zones, but they also fan out above the polar easterlies and the trade winds like the top of an anvil. Within them lie the jet streams. Over the northern hemisphere, the southern edge of the westerlies can be traced by following the path of the polar-front jet stream on weather maps.

The westerlies are the prime weather movers in the world, flowing predictably around the globe, dipping north and south in undulating waves that tend to pass over the same regions summer after summer and winter after winter. But occasionally they form a series of odd, elongated loops, leaving northeastern North America cold and rainy in July or central Europe balmy in December. No one knows exactly why such loops develop.

Around 6500 B.C., writes the British climatologist Hubert Lamb, the high-altitude westerlies in the northern hemisphere must have developed a bulge that sent them careening far south over North America, while over Europe the flow of air must have veered back north toward the pole. When such a trough develops, it bends the westerlies out of their usual, undulating pattern. West of the trough, polar air is funneled toward the equator; east of it, tropical air flows toward the pole. And, indeed, pollen records show that at the time most of North America was still buried beneath the glaciers of the Wisconsin Ice Age, while Europe was beginning to thaw. The ice covering North America was five times thicker than that covering Europe, so it took much longer to melt away. But Lamb notes that strong southwest winds, possibly caused by the presence of a trough, bore pollen all the way from what is now the southern United States to Greenland and Iceland.

Another noted climatologist, Reid Bryson of the University of

Wisconsin at Madison, has developed a computer model for tracking how circulation patterns have aided the ascent and heralded the doom of civilizations around the world. In the early 1960s Bryson and several students traveled to the wilderness of the Northwest Territories to study the freezing patterns of northern lakes. One day Bryson noticed an odd line of soil on a cliff at the edge of a lake. It was a fossil soil, laid down by a boreal forest that had once covered the countryside. "I'd been on that lake a number of times before and had never noticed that soil," he says. "I suppose I wasn't surprised to find it — but I was delighted."

Bryson and his students went on to find similar deposits over about fifty square miles. Later Bryson and several collaborators used radiocarbon dating to age soils from more than 150,000 square miles of tundra and trace the expansion and contraction of the forest over a 3,000-year period.

Just before 3000 B.C., coniferous trees grew far into the Northwest Territories, reaching nearly two hundred miles north of their present limit in northern Manitoba. A few hundred years later, the trees in the north abruptly died off. The forest limit ebbed and flowed a bit in the following centuries, reaching far north again by A.D. 1100. Then around 1200 it retreated quickly south, as if the realm of arctic air had suddenly expanded.

The decades around 1200 were a period of upheaval among the native peoples of the Great Plains. Scores of small villages east of the Rockies were abandoned; by 1300 the great settlements of the Southwest at Mesa Verde and Chaco Canyon had also vanished. All had depended heavily on corn for food. Might the civilizations have been destroyed by a prolonged drought, Bryson wondered, brought by a change in wind patterns? He and anthropologist David Baerreis conducted a study of Great Plains wind flow and village occupation rates during the twelfth and thirteenth centuries. Using the boreal forest line as a starting point, the two compiled a map of likely circulation patterns over the continent. A southward drift of arctic air would have strengthened west winds over the Great Plains, extending the rain shadow cast by the Rocky Mountains and causing a widespread drought.

Bryson and Baerreis discovered firm evidence of a drought in the fossil pollens and garbage middens of the Mill Creek culture of northern Iowa. At the remains of three villages they documented a rapid and drastic change in diet. Deer bones were abundant in layers of garbage dated to about A.D. 900. So also were shards from pottery vessels, which the tribe presumably used to store corn. Oaks grew in the area, and later willows. But as the drought took hold, hardy prairie grasses succeeded the trees. Deer bones vanished from the garbage pits, replaced by larger percentages of bison bones. By 1300 the number of animal bones and potsherds had dwindled. By 1400 the settlements had been abandoned.

In a book written with Thomas Murray, *Climates of Hunger,* Bryson notes that the centuries just before the disappearance of native farmers from the Great Plains must have been mild ones in the North Atlantic and Europe. The North Atlantic would have been on the east side of the atmospheric trough, with warm southerly winds flowing far to the north. The Irish and the Norse settled Iceland during this period, and around 982 Eric the Red scouted the coast of the island he named Greenland. Vineyards sprang up in England, a countryside that was formerly (and is presently) too cold for winemaking. But the mild weather was fleeting. The last settlements in Greenland died out in the 1400s, and excavations conducted after World War I uncovered bodies of dwarflike people, twisted and diseased from malnutrition. The atmospheric trough had vanished, Bryson and Murray write, and the northern climate had once again grown cold.

THE SEVERE cooling of the European climate in the fourteenth century began with a suddenness that shook medieval society. By the close of the thirteenth century much colder temperatures in Arctic waters had set up a thermal gradient that spawned extremely stormy conditions in the North Sea and the northern Atlantic, and catastrophic floods killed thousands of people along the coasts of Denmark, Holland, and Germany. Storms swallowed whole islands and carved out the Zuider Zee. In 1315 the grain harvest failed to ripen all across Europe, the first season of a 175-year stretch of damp, bone-chilling weather. Famine followed the failure of the grain harvest. Whole herds of cattle and sheep died in epidemics of diseases that incubated in the flooded landscapes. A poisonous blight blackened the kernels of rye. Ingesting even a single seed of tainted grain caused a terrible sickness known as St. Anthony's fire, which withered one's limbs and turned them black. In 1348 the bubonic plague appeared.

We will never know to what extent patterns of atmospheric circulation enhanced the climatic shifts that led to such massive suffering. Temperatures grew more moderate around 1500, but fifty years later another cold phase began. Judging by the sixteenth-century weather reports, tax records, and church records that have been preserved, climatologists suspect that around 1550 the westerlies began an unusual looping flow (by modern standards) that funneled arctic air over Europe. The result was the Little Ice Age, lasting roughly from 1550 until 1850. In the worst years the Scandinavian countries, the British Isles, and Europe experienced some of the coldest temperatures on record, while Siberia — in the warm southerly winds east of the trough — enjoyed periods of great warmth.

Hubert Lamb concludes from historical records that during the Little Ice Age the surface temperature of the North Atlantic off Greenland was nine degrees colder than it has been in modern times. As a result, severe cyclonic storms formed that far exceeded even the worst tempests of the twentieth century. Much of the destruction was caused by marauding sand that blew inland from the coast. Whole countrysides in Scotland and the Netherlands were consumed; one storm in 1697 utterly buried a four-thousand-year-old settlement in the Hebrides. In Denmark rolling pastureland became a dune-filled desert within a few weeks.

Cold water from the Arctic spread across the Norwegian Sea, bringing devastation to fisheries. Each year snow fell heavily in some parts of Europe, while other regions received much less. The variation, Lamb writes, came from the year-to-year position of the cold trough that pushed southward from the Arctic. Different countries suffered severe cold snaps each year as the trough bulged over various parts of the northern hemisphere. Switzerland recorded its most extreme temperatures in the winter of 1684–85, a year later than England. China, like Europe, suffered a period of devastatingly hard winters in the seventeenth century. But the cold came in waves up to twenty years later than it had come to the countries of Europe, as if the ice-bearing trough had slowly drifted to the east.

While the lands to the north were under the grip of killing cold, the Sahel region of western Africa was experiencing severe drought, apparently because the moisture-bearing monsoon does not migrate as far from the equator in cold periods as in warm. The monsoons of the world float northward and southward in concert with the Intertropical Convergence Zone (known popularly as the doldrums), where the trade winds of the northern and southern hemispheres meet.

In recent times the convergence zone has tended to drift as far into the northern hemisphere as 20 degrees. But in the depths of the Little Ice Age, it seems, the odd atmospheric circulation kept the convergence zone from making its normal northward trip, especially due south of Europe. Historical records show frequent failures and interruptions of both the African and Indian monsoons.

During the same period scientists believe that Antarctica experienced a much milder climate than it has had in modern times. I find it interesting that the bitter cold of much of the northern hemisphere was likely matched by comparative balminess in the far south, as if the earth fosters some sort of global symmetry, a climatic yin and yang.

IN MIDSUMMER the ocean off the northern Outer Banks warms and cools with shifts of the prevailing northeast-southwest wind. One July day I quit work early and head for the beach, where I expect to find pleasant blue-green surf, tousled gently against the shore by a light easterly breeze. Sure enough, the ocean is so clear and beautiful I can see sandbars shining through. I jump in and float on my back, bathed in the warm surface waters of the eastern Atlantic.

We live at the junction of two oceans, one of liquid, one of air. The currents that gently twist my body are not unlike those that fill the atmosphere, responding to the tug of cold and warmth. It is easy to forget that air is a fluid and behaves as water does, curling into wrists and fingers, ripples and waves. Were there a southwest wind today, the clear surface waters would have been pushed away from the beach, and bracing, muddy waters from forty feet down would have been sucked up to shore. In the same way, a strong wind might stave off a warm front and leave a region shivering beneath unseasonable cold. Air and water, two great, encircling seas, similar in form, each shaping the other's flow.

More than four hundred years ago, in July 1587, a favorable wind brought three ships bearing the first settlers from England to this section of coastline in what was for them a New World. The colonists who disembarked at a fort previously established by British soldiers on the north end of Roanoke Island hoped to establish a thriving settlement in the land they called Virginia. To reach the Outer Banks they had caught the same winds and currents that brought Columbus to the West Indies. Whether Columbus was an extraordinarily adept captain or an extraordinarily lucky one, the route he chose turned out to be the easiest passage, both across the Atlantic to North America on the trade winds and back to Europe on the midlatitude westerlies.

The 150-plus men, women, and children who landed on Roanoke found the fort demolished and set about repairing the settlement's few houses. No doubt daunted by the thought that Indians had massacred the soldiers, they nonetheless began carving out a "Cittie of Raleigh" in the dense, swampy forest. A month later the colony's governor, John White, returned to England for supplies.

Largely because of England's war with Spain, three years passed before White could make his way back to Roanoke Island, and then only as a passenger on a fleet of three ships. He found the settlement gone. Carved in a tree near the water was a clue to its fate, the letters "CRO." A tree near the fort bore the word "CROATOAN." White interpreted the carvings to mean that the settlers had gone to live among the friendly Indians on Croatoan, the name for present-day Hatteras

Island. He begged the captains of the expedition to sail south to look for signs of a village. But before a course could be charted, a strong wind sprang up, severing the anchor cable of one ship. When the storm abated, the captains agreed they could waste no more time and turned toward England.

Wind dictated that the first English colony would be established in the latitudes of the mid-Atlantic, and then storm winds kept the settlers (if they were still alive) from being rescued. No trace was ever found of the people whom White had left behind. The English did not gain a toehold on the continent until 1607, with the founding of Jamestown on the Chesapeake Bay. The Outer Banks remained unsettled until the 1680s. But what might have happened if White had located the settlers among the Indians on Croatoan? How might the history of this region, and colonial America, have been redrawn?

FOR THOUSANDS of years seafarers depended on wind to carry them to the outer reaches of their worlds. Although there is no clear evidence of when the sail came into wide use, archaeologists believe it originated on the Nile River and the Persian Gulf, perhaps as early as the fourth millennium B.C. Nubian rock carvings from that period include numerous images of reed boats with elaborate curved hulls and sails mounted in the center. If the carvings are accurate representations, the boats were large enough to carry livestock and up to fifty people. At first sails were probably made of skins and, later, of reeds woven into mats and held erect between two staffs.

By the third millennium B.C. sailing ships had appeared in the Mediterranean. A scribe's note dating from about 2650 B.C. tells of the arrival of forty ships bearing one hundred cubits of wood from the famous cedars of Lebanon. Early watercraft depended mostly on power from rowing; their simple square sails could catch wind only from the rear. The lateen, or triangular, sail did not appear until much later. Some historians believe it came into use on small ships as early as the Roman Empire, in the first century B.C. Even so, it was at least a thousand years before seamen learned to rig large vessels with sails that enabled them to travel with equal reliability in all directions.

Even the most skilled sailors found it difficult to buck prevailing currents and winds. In the first century A.D., the Greeks learned a secret that traders from India had known for centuries: seasonal monsoons could take them east across the Arabian Sea in winter and back west in summer. The discovery at last provided the Greeks with a way of reaching the spice islands of the Far East. From the seventh through the ninth century, Norse explorers settled the North Atlantic, sailing to the new lands on the light easterlies encircling the poles, moving from island to island and perhaps to the North American continent. But in the fourteenth century, as the climate cooled, the sea became increasingly stormy and filled with ice. The northern sea route was abandoned, along with the connecting villages along it.

When Europeans attempted to sail south down the west coast of Africa, they were repelled again and again by northbound currents and winds. The Portuguese finally rounded the Cape of Good Hope in 1487, then quickly returned home. When Vasco da Gama made his historic voyage of 1498, his crew found the people of the east African coast accustomed to wearing fine silks and trading with travelers from distant lands; local kings scoffed at the cheap beads da Gama offered in hopes of winning their friendship. The Chinese had been journeying to those parts on favorable winds and currents for eighty years. After 1500 the standard route from Europe around the Cape of Good Hope took sailors thousands of miles out of their way. Instead of beating due south, ships would tack southwest from the Canary Islands almost to Brazil, then catch the roaring forties back to the east.

In all historical periods, as explorers fanned out across the oceans in search of new lands, they generally sailed easily from east to west in tropical latitudes. But when they tried to retrace their routes, their progress was labored. Spanish explorers venturing westward from Mexico and Central America made quick passage to the Philippines

— eight to ten weeks — but were unable to return. Those who tried to find a counter-wind on the edge of the trades spent weeks marooned in the horse latitudes, or doldrums. The only way to get back to the New World, it seemed, was to circumnavigate the globe and again approach it from the east. Finally, in 1564 and 1565, explorers discovered the great gyre that would carry them east across the Pacific. First they sailed north along the Asian coast with the Kuroshio Current. Then they rode the prevailing westerlies of the temperate latitudes to the West Coast of North America and beat their way south. The trip was difficult, especially the last leg in the unpredictable winds of the North American coastline, and required four to seven months.

Spanish explorers were astounded to discover that the indigenous peoples of the western South American coast were skilled sailors. The reed rafts known as balsas, equipped with steering rudders and large sails, had been in use for perhaps two thousand years. Some were seventy to eighty feet in length, as long as the European caravels. Both the South Pacific aboriginal peoples and the Chinese employed a system of centerboards or rudders that enabled them to tack close to the wind. The Polynesians used outriggers for balance on their canoes and a triangular sail rigged with the narrow tip at the bottom (upside down by modern standards). In his book *Early Man and the Ocean,* Thor Heyerdahl quotes the 1680 journal of a Spanish buccaneer as reporting that Incan balsa rafts sail "excellently well" and despairing that his ship would not be able to catch one plying a stiff offshore wind near the Galápagos Islands.

How extensively did the native peoples of the South Pacific explore surrounding oceans? Heyerdahl argues that the Incas and Polynesians, with their sophisticated rafts and canoes, probably ventured quite far, sailing west and north to the Aleutian Islands, which could have served as steppingstones to cultures in the West. (Other writers suggest that the ancient Chinese may have made the same trip.) He notes that tribes of the Malay Archipelago bear striking physical and cultural similarities to the peoples of the Brazilian jungle; even the poisons used in their blowpipes are closely related. In addition, studies show that the petroglyphs of the Pacific Northwest resemble those of the Hawaiian Islands, suggesting that at some point the two cultures were in close contact. And early in the twentieth century, Heyerdahl writes, hundreds of redwood logs from the rivers of the Northwest drifted to the Hawaiian coast. Why couldn't the vessels of aboriginal travelers have made the same trip?

Persuasive though they may be, Heyerdahl's arguments are disputed by archaeological findings that suggest the Polynesian aboriginals migrated eastward. Beginning about three thousand years ago, archaeologists believe, the Lapita peoples journeyed from the region of the Solomon Islands 250 miles east to the Santa Cruz Islands, then southward to the New Hebrides and eastward again more than 500 miles to Fiji, Tonga, and Samoa. If they did, their sailing skills must have been all the more impressive to have bucked the prevailing currents and wind.

Whatever the course of their explorations, the Polynesian peoples were attuned to wind in ways that we can barely fathom. Many tribes attributed to breezes the power to carry myths, legends, news, and gossip about royalty. In the open sea helmsmen navigating without benefit of instruments used the feel of ocean swells, generated by prevailing global winds, to find their way from island to island. These subtle, rocking swells have more distance from crest to crest than surface waves kicked up by local winds, and they pass more slowly. Also, their uniform shape differs noticeably from the shapes of swells distorted by land. As secondary guideposts, the Polynesian navigators used the direction of the wind, the presence of phosphorescent plankton, the behavior of seabirds, and the appearance of clouds.

In the 1960s David Lewis, a British adventurer raised in the Cook Islands and New Zealand, studied traditional Polynesian navigation techniques and attempted to sail a catamaran without instruments from Tahiti to New Zealand, a distance of 2,200 miles. A skilled captain traveled with him but kept the charts and modern navigational equipment under lock and key. At one point the captain had to intervene to keep Lewis from missing the Cook Islands and sailing into open ocean. Except for that one error, Lewis reached New Zealand unaided, only twenty-six miles south of his intended destination.

Lewis came away from the experiment awed by the skill with which his native teachers could read the oceans. He writes of swells off the Santa Cruz Islands that come from the northwest and southeast, traveling for hundreds of miles and passing through each other "like the interlocked fingers of two hands." In the Carolines he asked one of his teachers to demonstrate how the swells of an unfamiliar region might be deciphered. "He studied them at frequent intervals for hours at a time, when necessary orientating them by the sun morning and evening, until he could recognize the shape and characteristic of each swell," Lewis writes. "Once they had been sorted out and mentally 'labeled,' the different swells appeared to become as recognisable to him as people's faces." Reading Lewis's account, I cannot help but be saddened by the thought that such ancient perceptions of wind, and of our world, are condemned to obscurity by our complete dependence on modern electronic equipment to help us chart our way.

KNOWLEDGE of wind patterns also gave early cultures an advantage in trade and war. In the fifteenth century B.C., at the height of the Egyptian empire, a wave of Indo-Europeans invaded the island of Crete and conquered the Minoans, setting the stage for the rise of the Greek culture celebrated in the epic poems of Homer. By all accounts the early Greeks were a sailing (and rowing) people, fond of adventure, and given to both vigorous trade and piracy. Historians believe that the Trojan War may very well have started not with the abduction of Queen Helen but with disputes over Mediterranean trade routes.

The first Western people to show true finesse in their use of the sail were the Phoenicians, who learned to navigate by the North Star and who ruled the Mediterranean between 1200 and 900 B.C. Rather than island-hopping, as the Greeks did, the Phoenicians struck out across the Mediterranean from their eastern strongholds to Iberia (now Spain and Portugal), sailing night and day, fetching home wheat, oil, wine, and tin, and colonizing the coast of North Africa. It is believed that they may have been the first Western people to sail around the southern tip of Africa.

For several centuries the trade routes of the Mediterranean were contested by a half-dozen cultures, among them the Phocaeans, the Etruscans, and the people of the easternmost Phoenician colonies, the Carthaginians. Near the end of the sixth century B.C. the Persians began a full-scale campaign to overrun the empire of Greece. They conquered the Aegean islands, but a sudden storm prevented the Persian navy from assaulting the mainland. Dozens of the invaders' ships were smashed against rocks.

Over the next two decades the two powers continued to joust for control of the region. In 480 B.C. Persia, having driven the Greeks and their allies to a final stronghold on the island of Salamis, gathered for a decisive battle. The Greek forces were surrounded and badly outnumbered; Athens and the Acropolis were in flames. With only 360 vessels, it seemed impossible that the Greeks could hold off the Persian navy of 1,400 much larger ships. Little did Xerxes, the Persian king, suspect that the balance of power hung on the direction of the morning wind.

The Greek strategy began with a ploy by the Athenian admiral Themistocles, who knew Xerxes to be arrogant and stubborn. Themistocles dispatched a messenger to "leak" to Xerxes the information that the Greeks planned to withdraw from Salamis under the cover of darkness. To succeed, the withdrawal would require the utmost secrecy. The Greeks had developed a narrow, low-slung ship called a trireme that employed nearly two hundred oars, but at night and without wind it could not move faster than four knots. In the narrow straits around Salamis the Greeks would need at least a six-hour lead to outrun the Persians. Otherwise they would be overtaken by the enemy's larger warships and slaughtered.

From accounts of the battle left to history by Herodotus, it seems that Xerxes never considered the possibility that such a jewel of strategic intelligence might be false. The bait must have been tempting. If he could entrap the fleeing Greeks, he could crush their whole navy with a single blow instead of fighting a series of protracted battles. He ordered ships to block the narrow Megara Channel on the west side of Salamis, eliminating one route of escape. Then the Persians and their allies — who normally did not even keep up regular patrols after dark — readied themselves for a night of war.

Midnight passed, and the wee hours. By dawn it must have occurred to Xerxes' captains that they had been deceived. But they lived in such fear of the king's wrath that no one dared suggest that something was amiss. Soon after sunrise Xerxes and his high command, comfortably seated on a mountainside that gave them full view of the coastal waters, saw the Greek fleet streaming in disorder to the north. The Persian command had expected a retreat to the south. Instead, it seemed that the Greeks intended to sail completely around the north end of Salamis and escape through the Megara Channel, as if they did not realize it had been blocked. At Xerxes' command most of the Persian navy gave chase.

It is safe to assume that none of the pursuers understood the significance of the band of clouds that foretold a building breeze. As the Persians entered the narrow strait on the east side of Salamis, the Greeks suddenly turned to fight. "An echoing sound of battle, like some triumph-song / Went up from each Greek throat," wrote the Greek dramatist Aeschylus.

By late morning the Etesian wind, a cool, gusty northwesterly, was roaring across the sea, generating swells of such size that the top-heavy Persian warships rolled and yawed. Though the vessels on the front line tried to back out to more open waters, behind them were scores of others whose captains were bent on sailing into the fray. The Athenians' slender triremes, nimble even in choppy water, began ramming the Persian vessels. Within a few hours the water was thick with wreckage and corpses. A "vast mass of Persian ships — many of them badly crippled, with trailing spars and cordage, oars broken off short, timbers sheared or sprung by those terrible bronze-sheathed battering rams, went streaming away," writes the historian Peter Green.

The Persians were vanquished, though the Greeks did not realize it for some time. They set about repairing their damaged ships, waiting for Xerxes to mount a new assault. But with the morale of his navy in tatters, Xerxes sailed for home. The freedom of Greece, writes Green, no longer stood "on the razor's edge. At the eleventh hour, and against all expectation, Greece had been saved, and not even his bitterest enemies — of whom there were many, both at home and abroad — could deny that it was Themistocles who had saved her." Themistocles — with help from the wind.

In the years since, many other winds have played decisive roles in battle or shaped the course of history in unexpected ways. Kublai Khan at last abandoned his quest to overrun Japan after his ships were deterred by a violent north gale in 1275 and, six years later, in 1281, by a typhoon. In 1524 strong southerly winds prevented the Spanish explorer Francisco Pizarro from sailing south along the Pacific coast of Central and South America. As a result, the explorer's backers questioned his credibility, and his campaign against the kingdom of the Incas was delayed for several years — during which time the last of the Incas' strong monarchs died and a rivalry between his sons thrust the empire into civil war. In 1588 the Spanish Armada, on its way to storm England, met with steady southwest winds that kept its heavy men-of-war from maneuvering as deftly as the swifter British warships. That, at least, is how legend describes the battle; the medals awarded to British naval officers bore the inscription, "God breathed and they were scattered." Careful historical accounts show that the Spanish were simply outmaneuvered and outgunned. However it happened, the Armada was disabled bit by bit until a wind shift to the north opened an avenue of escape. In the gales that followed, many Spanish ships foundered, and the English emerged decisively victorious.

Wind continued to shape history through the age of sail, as the Chinese plied the oceans with their nine-masted, 400-foot-long junks (in the early fifteenth century, the Ming dynasty raised a navy of 3,500 vessels) and the Europeans sailed cogs, caravels, and galleons. The perfection of the fully rigged ship sometime in the fifteenth century enabled boat-builders to craft much larger vessels and opened the seas to wider travel than ever before. English great ships carried two and three tiers of guns; sleek Dutch *fluits* maneuvered nimbly from port to port. In the 1840s the grand clipper ships began rounding South America's Cape Horn to bring back gold from California. During the same decade, however, the British inaugurated steamship service from Liverpool to North America, and the age of sail came to an abrupt close. A few sailing ships continued to travel between North America and Australia as late as the 1940s, but on the North Atlantic, steam quickly usurped the domain of sail.

Since that time, wind's influence on history has been more subtle. Yet even in the postindustrial age, winds can turn the tide of battle and the fortunes of people. Between 1932 and 1937, a strengthening of the westerlies over North America brought disaster to the farms of the Midwest. Dry winds from the Rockies swept the region, parching the land, plucking up soil once held in place by prairie grasses and carrying it east in boiling clouds. On May 12, 1934, the *New York Times* reported that dust from the plains hung thick in the atmosphere, throwing Manhattan into an odd midday twilight. Coinciding as it did with the Great Depression, the havoc wrought by wind left both physical and psychological marks on the nation.

During the North African campaign of World War II, Allied and German troops were several times forced to halt in mid-battle because of sandstorms caused by the khamsin, a hot desert wind that reaches speeds of ninety miles an hour. Grains of sand whirled by the wind blinded the soldiers and created electrical disturbances that rendered compasses useless. In May 1942 wind-created static electricity was blamed for an explosion that destroyed an Allied ammunition dump in eastern Libya. Later, the D-day invasion was postponed because of foul weather, and troops waiting to attack the coast of Normandy became badly seasick as they rode over towering waves stirred up by storm winds. Although the invasion was successful, the troops' weakened condition and the lack of maneuverability caused by the steep swells were partly to blame for the high death toll that day.

In the late 1960s, the westerlies of the northern hemisphere strengthened and moved south over Africa, holding the moisture-bearing monsoons farther south and parching countries along the continent's girth from Senegal to Ethiopia. Like the land-sea breezes familiar to seashore visitors, monsoons are fueled by differences in temperature between earth and ocean. As the Intertropical Convergence Zone wanders away from the equator, its clouds gather moisture over the ocean. When these clouds pass over Asia, India, and Africa, the moisture falls as soaking seasonal rains. But if the zone fails to follow its usual migration patterns, millions of people suffer. At the close of the 1960s the dry, hot weather characteristic of the Sahara began creeping southward into the semiarid Sahel. Drought years followed one upon another; 100,000 people died of famine. Only massive shipments of grain from other countries kept the death count from climbing much higher.

Wind delivered a memorable coup de grâce in May 1980, during an attempt to rescue the hostages at the American embassy in Iran. On that night seven Sea Stallion helicopters were bound for a base camp deep in the Iranian desert. An eighth helicoptor had been grounded by an equipment problem. The remaining pilots, flying under cover of darkness, could not risk alerting the enemy by talking over their radios; they planned to keep each other in sight. But less than halfway to their destination their vision was completely blocked by the airborne dust of a severe sandstorm. They flew out of one cloud and immediately encountered another. When the gyroscope failed on one craft, its dizzy pilot turned back. Then another helicopter developed a problem with a hydraulic pump. To attempt the rescue with only five helicopters was far too dangerous; the mission was called off. During refueling operations for the return flight, two helicopters collided and eight men were killed.

Later much would be made of the strike force's lack of preparation for the mission and of Jimmy Carter's ineptness as commander in chief. But had that sandstorm not caused the second helicopter to turn back, the mission would have proceeded, and the hostages might have been freed. If they had, the election of 1980 and the political events of the ensuing decade might have taken a radically different course.

IT IS ANOTHER mild summer day on the Outer Banks, with no threat of storms or climatic change. I stand on a dune in a light northeast wind, the ocean behind me, trying to make sense of an old meteorological tenet known as Buys Ballot's law, which says that if the wind is at your back, high pressure will be to your right. This holds true only in the northern hemisphere, where fair-weather winds move clockwise. In the southern hemisphere, I would have to look to the left.

High pressure to my right, to the west. Good, I think. Nice weather for the coming weekend. I see, and foresee, nothing but blue skies. A change will be due when the winds drift to the southeast and the high pressure passes to the north and then offshore. Again the Outer Banks breezes appear to me as a dragon. This time the great beast is circling, circling, perhaps beating down the grasses, preparing a place to lie down. If it ever backs up — counterclockwise, cyclonic, as suddenly as if it has stepped on a thorn — I know there will be hell to pay in terms of weather.

Island lore warned that backing winds were full of menace long before scientists explained that they belonged to the circulation patterns of low pressure. I like the idea that there is value in both ancient and modern characterizations of wind, and in this I am not alone.

Early in the history of wind and weather science, Robert FitzRoy, the British captain who piloted the *Beagle,* which ferried Charles Darwin around the world, urged that scientific observation and traditional forecasting methods be meshed. "The more carefully we compare and combine forecasting material obtained from two different sources, the more satisfactory will be the results," he wrote in *The Weather Book,* a noted work published in the mid-1860s in both English and Russian. An adept navigator, self-educated in the nuances of weather, FitzRoy was so skilled at foretelling storms and heavy winds that English and Scottish fishermen came to depend on his predictions.

In 1805 Commander (later Admiral) Francis Beaufort of the British navy devised the Beaufort wind force scale, based on the stress exerted by winds of different strengths on a fully rigged man-of-war. Beaufort's original scale contained twelve ratings, or forces, from calm air to hurricanes. It made no reference to wind speed. Later the scale was expanded to include observations of sea states. A strong breeze, force 6, is one in which large waves begin to form, whitecaps are common, and the topsails must be set at the first reef. In a whole gale, force 10, the wind whips up high waves with long overhanging crests, the surface of the sea appears white from blowing foam, and a ship can keep only its reefed main topsail and foresail up. A storm is considered to have reached hurricane strength when the air is filled with spray and a ship can carry no sails at all.

In 1874 the International Meteorological Committee adopted the Beaufort scale as a standard for weather forecasting. However, physicists did not work out accurate equations for correlating force numbers to wind speeds until the early years of the twentieth century. The current Beaufort scale, which assigns specific speeds to each force, was not adopted internationally until 1939. As anemometers became more precise, the Beaufort scale fell out of use by meteorologists.

The mid-nineteenth century was in many ways the genesis of modern wind forecasting. Until then it had been difficult to measure the speeds of winds, much less predict them. In a simple cup anemometer, the revolving cups operate a counter; such devices had been used in the Orient for thousands of years to estimate wind speed. But it was not until 1850 that an Irish physicist named T. R. Robinson published a simplistic and, as it turned out, inaccurate theory on how the instrument might be calibrated. Six years later an American, William Ferrel, published *Essays on the Winds and Currents,* an important work in which he correctly explained how wind is deflected by the Coriolis effect.

The ships' captains who sailed the well-traveled routes of trade had abundant knowledge of regional wind patterns, but scientists had never attempted to correlate the seafarers' expertise. In 1842 an American naval officer, Matthew Fontaine Maury, published a circular urging that at the end of each voyage captains send their logbooks to the U.S. Depot of Charts and Instruments in Washington so he could comb through them for information about currents and winds. Maury hoped to compile an atlas of the world's winds, storm systems, and ocean currents — a mammoth undertaking — but initially few captains complied with his request. Disappointed, Maury concentrated on a single route, the passage from New York to Rio de Janeiro. He analyzed the scant records available and drew up a course that, he said, would enable ships to make optimum use of currents and winds. Shortly thereafter a captain reported that by following Maury's course he had made the trip south in twenty-four hours instead of the accustomed forty-one hours. As word spread, Maury started receiving dozens of logbooks each month. Beginning in 1851 he published a series of charts that brought him international renown. By following his suggested routes, ships were able to sail from New York to California in 135 days instead of 180, and from England to Sydney, Australia, in 130 days instead of 250 days. Maury drew one of the first accurate maps of global winds and currents and won international acclaim.

By the late nineteenth century it was well known that foul weather was associated with low barometric pressure. Meteorologists learned to draw lines of equal atmospheric pressure — isobars — on weather maps to chart the movements of weather systems and watch them change character. Barometric readings were passed from station to station by telegraph. Even so, meteorologists could make only the most rudimentary forecasts: on the following day the weather in a region would probably be dry, or wet, or stormy.

In the early twentieth century, scientists began experimenting with weather balloons and so learned that wind speeds increase with height. The aerologists, as they called themselves, also discovered that sometimes winds at the surface blow from a different direction than do winds at a thousand feet or higher. The researchers found a zone of air surrounding the earth — a planetary boundary layer — where tremendous turbulence is generated as air passes over mountains, plains, and cities and as convection sends hot currents roiling skyward. The boundary layer is analogous to the bottom layer of water in the ocean, where intricate eddies form as currents pass over canyons and reefs. Contained wholly within the lower troposphere, it varies in thickness from a mile to three miles, depending on the surrounding terrain. As far as the aerologists could tell, from the boundary layer to the edge of earth's atmosphere the flow of air was smooth.

The study of wind took on new urgency with the increasing use of airplanes and the outbreak of World War I. Suddenly there was an urgent need to know when planes could be flown, how bombs would drop, when gas attacks could be waged. But even as knowledge of atmospheric flow patterns grew, the behavior of low-pressure cores continued to defy explanation. Into the 1920s forecasters relied more on statistical patterns than on actual observations of atmospheric conditions. The science of meteorology was still in its infancy; no one really understood how weather systems formed and advanced.

On the morning of October 22, 1921, barometric readings alerted the forecaster at the Danish Meteorological Institute in Copenhagen to an approaching low-pressure system. He assumed it would pass quickly to the east and predicted that the strong winds of the morning would lessen by afternoon. Gale warnings were suspended. And, indeed, in Copenhagen the sun came out and the heavy winds ceased. In northern Denmark, however, a storm with hurricane-strength winds pressed hard against the coast, sinking dozens of ships and swamping villages. As the winds intensified, all lines of communication were broken. The forecaster did not learn of the devastation until the next day.

The Danish forecaster's mistake might have been understandable, given the crude state of the science, except that meteorologists in Norway and Sweden had correctly predicted the powerful storm. The Norwegians and Swedes were versed in a new technique devised by a group of forecasters under the leadership of the Norwegian physicist Vilhelm Bjerknes. Bjerknes was the first to recognize that weather is a continuing march of air masses and fronts. He and his colleagues also developed a new model of the midlatitude low-pressure systems that meteorologists call extratropical cyclones. Their theories revolutionized the science of weather forecasting.

Bjerknes's approach enabled meteorologists for the first time to study the formation of pressure systems and predict their movements and changing shapes. By tracking fronts, they could also predict the direction and gustiness of the wind. Yet they still did not have the skill to analyze global weather patterns, and their forecasts were vague and frequently inaccurate. Nor did they know much about upper-level atmospheric circulation. In 1928 the Swedish meteorologist Tor Bergeron suggested that three parallel cells, or looping belts of wind, circulated in both the northern and the southern hemisphere. Bergeron believed that the movement of surface air was countered at higher altitudes by a flow in the opposite direction. According to his model, tropical surface winds flowed east, rose in height, and flowed back west. The same thing happened at the poles. In temperate latitudes the surface westerlies were overlaid with strong easterly winds. These great rotating cells of wind kept the atmosphere constantly turning and constantly balanced. Bergeron's so-called tricellular theory was simple, easily understandable, and broadly embraced. It was also wrong, as events would demonstrate dramatically during World War II.

In May 1945 the climatologist Reid Bryson was a captain with the Army Air Corps in Saipan. One day, just before a planned mission to bomb military installations on the Japanese coast, his colleague William Plumley was asked to devise a prediction of the wind speed at 33,000 feet for the following day. Plumley and Bryson had done similar calculations in Hawaii and — counter to tricellular theory — they strongly suspected that the high-altitude winds would be flowing from the west. They examined known atmospheric flow patterns, noted the presence of an approaching cold front, calculated the differences in temperature at various altitudes, and came back with a predicted westerly wind speed of 168 knots (193 miles an hour). "In five minutes," Bryson says, "the general was over at our office, chewing us out and questioning our ancestry because of our stupidity. He told us to refigure our calculations." Again Plumley and Bryson came up with westerly winds of 168 knots. The following day, pilots flying at 33,000 feet on their way to Japan encountered a severe flow from the west. They clocked it at 170 knots. "The general was a gentleman," Bryson says, "and he apologized to us." In the end the upper-level westerly winds both debunked tricellular theory and forced the American High Command to abandon its strategy of bombing military installations from above 30,000 feet. Instead pilots ran low-level incendiary bombing raids on Japanese cities under the command of General Curtis LeMay.

In the last years of World War II, Allied pilots, flying high to stay out of range of antiaircraft guns, discovered what the Japanese had encountered earlier in the war: the rushing rivers of air now known as the jet streams. In 1944 and 1945 the Japanese launched hundreds of hydrogen balloons, unmanned but armed with bombs, which the jet stream carried to North America — 5,000 miles — in three days. One reached the coast of Oregon and exploded near a Sunday school picnic, killing five children and the minister's wife. Another descended near Hanover, Washington, landed on a power line, and shut down the reactor that was manufacturing plutonium for the nuclear bomb to be dropped on Nagasaki. Other balloons were found in the ocean, in Alaska, in Arizona, Montana, Kansas, Saskatchewan. One reached Farmington, Michigan, just outside Detroit. For the most part they caused no harm. But the potential for psychological damage was so great — enemy bombs drifting on the wind! — that the War Department ordered reports of the balloons kept from the public.

The Japanese sent the balloons via the polar-front jet stream, the great flow that steers weather systems across the American continent and Europe. After the war an intense international probing of the atmosphere discovered four separate jet-stream systems in the northern hemisphere — the polar front, a subtropical jet, and a polar vortex, all flowing from the west, and an easterly tropical jet. Each is thousands of miles long, more than a hundred miles wide, and several miles deep. They can occasionally reach speeds of nearly four hundred miles an hour.

Where the jet streams wander depends on the difference between temperatures along the equator and at the poles. In winter, when the temperature contrast is greatest, they drift closer to the equator. The subtropical jet forms and flows eastward, dancing with the polar-front jet, moving close and then away. Over North America, Europe, and Japan, the two systems sometimes intermingle, forming a vast, swift current that begets torrential rains and howling blizzards. At the same time, the polar vortex forms and flows eastward around the pole. These systems are mirrored in the southern hemisphere. Over Antarctica, the polar vortex encircles a body of air where the protective layer of ozone has become alarmingly thin. This hole in the ozone forms each winter when the polar vortex prevents air over the South Pole from mixing with the rest of the atmosphere.

In summer the polar vortex and subtropical jets disband, leaving a void that is filled, in the northern hemisphere, by the easterly tropical jet over India and equatorial Africa.

On the margins of the jet streams are extreme wind shears, where the speed of flow changes abruptly within a few feet. These can create a violent roiling of currents known as clear-air turbulence. Closer to the ground, where deserts and fields and cities concentrate heat, convection sends geysers of air shooting upward, past the invisible skin of the boundary layer. The smooth flow that early meteorologists envisioned for the upper atmosphere does not exist. Instead the air above us is filled with waves and crashing breakers, their crests sometimes marked by the formation of small clouds; with wide, smooth currents that stretch sheets of clouds into rippled ridges; with layers of air flowing in different directions that pull vapors from plummeting ice crystals to form mare's tails. Watch the clouds; they are evidence of our atmosphere's complex restlessness. They are all that can be seen of earth's invisible, foamless sea.

THE WIND, the wind. I hear it comb the pines above our house each day when I awake. After breakfast I listen to our dilapidated weather radio in an attempt to discern its shape. "Good morning," says the announcer matter-of-factly, "here's the eight A.M. weather roundup for the region. Norfolk, clear, winds north at fifteen knots. Elizabeth City, partly cloudy, winds north at ten." The winds are given first from stations that lie to the north, then to the south: Duck, northeast at fifteen. Manteo, northeast at fifteen. Frisco, east at twelve. Hatteras Inlet, southeast at ten. And on, until I know the dragon's path for the day.

I think of how much progress has been made in charting the wind, and I am amazed. Wind and weather have given up so many of their secrets that even veteran forecasters wonder how much more will be revealed.

A few months ago I went to Wolfeboro, New Hampshire, to visit meteorologist Bob Rice at his private forecasting center, Weather Window. I first heard of Rice's prowess at forecasting wind from some champion hot-air and helium balloon racers in Albuquerque. In 1978 he served as a weather adviser for Maxie Anderson's *Double Eagle II,* the first manned balloon ever to cross the Atlantic. I had spoken to him several times by phone. "When you stick a balloon in the air," he is fond of saying, "it acts just like an air particle. Using a balloon is a good way of finding what's happening aloft. The key in racing, of course, is to try and figure out what's happening up there *before* you launch the balloon."

Rice is famous among competitive sailors as well as balloonists. In 1995 he spent seven months in San Diego working with Team New Zealand in their successful bid to capture the America's Cup. During my visit to the offices of Weather Window, a half-dozen calls came in from competitive sailors in France and England.

Rice has thick ivory hair, a white, cropped beard, and a deep, rich voice. He and I chatted briefly about the odd weather of the summer. The East Coast had been unusually cool and rainy, but a prolonged drought was wreaking havoc in the ranch country of Texas, Arkansas, and Oklahoma. A long, south-reaching loop of westerly wind was trapping dry air in the western half of the continent and funneling rain in great bursts to the East. "It's not that unusual for a wave to form in that position," Rice said, "but the amplitude *is* unusual." He mentioned that a similar loop had developed over the Rocky Mountains in 1993, when the Midwest was inundated with rain and thousands of homes in the Mississippi watershed were lost to flood. "There was a long wave trough over the mountains, and the jet stream funneled through there so that cold air suddenly met a mass of warm air. Thunderstorms galore. Those things just kept firing through, one after the other. The ground couldn't absorb it that fast."

Rice was introduced to meteorology in the air force during the Korean War. "Prior to that," he said, "I hadn't given any more thought to weather than the next person. What captured me was the idea of weather as a living, breathing thing. You don't just type in a bunch of numbers and data. You visualize it. You see waves, you see storm clouds, you see the system."

Over forty years Rice has watched weather forecasting methods become increasingly sophisticated. In the early 1970s the entire emphasis was on the next day's weather. "It took us so long to build a twenty-four-hour forecast," he said, "that we almost didn't have time to do a forty-eight-hour forecast. By the time we got to the seventy-two-hour forecast, we might just as well have been throwing darts at a board. Airflow over terrain can be so complex. Every little twist causes a change. Now I just stand in awe of some of the models that are in use."

In 1994 Rice guided the *Enza*, a ninety-foot catamaran, in a quest by Captain Peter Blake to capture the Jules Verne Trophy for the fastest around-the-world sailing time. He stayed in touch with Blake by phone and fax. When the *Enza* reached the southern hemisphere, Rice hoped to keep her in moderate west winds on the northern edge of the roaring forties. But as the boat neared New Zealand, a computer simulation revealed the birth of a low-pressure system that threatened to set up easterly gales. "I saw it developing six or seven days in advance. I had time to get the boat farther south and keep it in the west winds. The point is, fifteen years earlier I wouldn't have had any idea that cell was developing. Not a week ahead of time. I might have had two days' warning, and that wouldn't have been enough." Thanks largely to Rice's guidance, Blake finished the trip in just under seventy-five days and captured the trophy.

The great progress made in improving forecasts did not continue into the 1990s. "Things have kind of plateaued for the moment," Rice said. "There are a lot of models out there, and half are very good and half are bad. We've got some big questions left to answer, and I'm not sure how well they can be answered. What causes the high-amplitude [looping] waves? I'm not sure anyone knows." El Niño, the warm pool of water that develops unpredictably in the South Pacific, may play a part. At the very least it must help shape atmospheric circulation. But, Rice said, "I hate anything that's used for an explanation for everything — heat, cold, drought, acne, smelly feet — the way El Niño is. There are so many interlocking factors. I don't think any one thing can be the cause."

HUGE GAPS remain in our knowledge of wind at all altitudes. During the mid-1990s the National Weather Service completed the installation of a Doppler radar system that can draw profiles of the winds across the continent. The radar tracks the movements of raindrops, insects, and dust to reveal the movement of air. It is capable of spotting the signature of a tornado fifteen minutes before it touches down. But it scans the horizon from fixed points, slanting upward to heights of 50,000 and 60,000 feet, so it cannot see the surface gusts and eddies, the isolated squalls, that may be developing over a particular locale.

There are also complex winds at the outer rim of our atmosphere, still so little understood. Near the equator lies an odd belt in which the winds of the stratosphere blow around the globe from the east for between twelve and fifteen months. Then they abruptly shift and blow from the west for thirteen to sixteen months. This great wind reversal, called the quasi-biennial oscillation, was discovered only in the mid-1960s. Scientists are slowly exploring the oscillation to see how it might affect the weather of the world. It is an odd phenomenon, apparently driven by atmospheric waves that are set up by convection.

The quasi-biennial oscillation may play a pivotal role in determining the intensity of hurricanes, those marauding tempests whose paths are so difficult to predict. Since 1983 meteorologists at Colorado State University, led by William Gray, have been honing a model for hurricane prediction using sea-surface temperature and pressure, rainfall in the Sahel and Gulf of Guinea regions of West Africa, the quasi-biennial oscillation, the presence or absence of El Niño, and other factors. The model uses statistics about atmospheric conditions in past years to predict what will happen next season. Gray and his colleagues can say that in years with high rainfall in the western Sahel and the Gulf of Guinea, major hurricanes are more likely to form, but they can only speculate as to why. (Neither can they predict how many storms in a season will make landfall and become killers.) They do know that fewer very intense hurricanes form when the oscillation flows from the east. In the easterly phase, an intense wind shear develops that acts like a huge saw, cutting the tops off tropical storms.

Perhaps in the future the secret of hurricane formation will fully yield to the scrutiny of atmospheric scientists. Perhaps in time we will also be able to predict when an odd looping pattern will develop in the storm-steering, high-altitude westerlies and bring weather considered, in our mild age, far out of the norm. Or perhaps we will never learn much more than we now know. William Gray, like Bob Rice, is skeptical of forecasting climatic trends using past weather and wind patterns. "The atmosphere is much too complex to model with computers," Gray told me when I went to visit him at his lab. "I don't think we'll ever be able to do it. I think it's arrogant to try."

Others, however, do believe that atmospheric modeling is possible and that in time we will be able to foretell with precision, at least a few years in advance, what wind and weather and climate will do, and why. Perhaps one day we will even be able to predict when the earth will enter another ice age, or whether it will warm, instead, like the interior of a greenhouse. But I stand with the Rices and Grays of the world. In the end I think we will discover that it cannot be done.

The wind is not a simple heat engine, after all, but a living organism with many members and many personas. I take solace in that knowledge. I take heart in the notion that it cannot be shackled, physically or intellectually, that it will always retain an immense capacity to baffle and surprise.

FOUR

Masters of the Breeze

I AM DRIVING south as fast as I dare, racing a rosy light to the north end of Hatteras Island. It is 6:45 on an early October morning. Last night a cold front, the first strong breath of winter, bore down on the Outer Banks, making its presence known with a bracing northeast wind. All thoughts of a lazy, lingering summer have been banished; the animals of the region are up and on the move. Another moderate northeast breeze is forecast for today, and I am on my way to a wildlife preserve to watch for the birds of prey that will almost certainly stream by.

I cross the curving bridge over Oregon Inlet without slowing to admire the fluid dance of ivory sand and cerulean surf. The sooner I reach my destination, the more birds I am likely to see. The bridge dumps me onto the spine of the island just west of a line of high dunes. Another four miles and I will reach a trail that leads to an observation platform deep in the Pea Island National Wildlife Refuge.

This stretch of road is a famous speed trap, so I hunch over the wheel and focus on keeping the speedometer a bit below sixty-five. I think of the falcons I may be missing and curse my early sluggishness, my theft of an extra half hour's sleep. The road has become a tunnel; nothing exists outside its narrow bounds. But just as I reach a long straightaway a blue-black bullet causes me to brake. A peregrine falcon swoops low over my truck and disappears on the ocean side of the dunes.

For a moment I sit motionless in the road. Coming to my senses, I gun the engine and follow the falcon south, reaching a speed of seventy-five before I spot it skimming the ridge to my left. I slow to fifty and keep pace, floating with it as if airborne myself, cherishing the sight of its executioner's mask and spear-tipped wings. A half-mile south the dunes rise to a crumbling wall and divide us.

From behind a cloud the sun emerges into a pale green sky. A gilding light slides across the sandy meadows and marsh. I let out a breath, push back in my seat, and coast.

A SHORT TIME later I stand on the observation platform in full sun. The cold front, offshore now, has lessened its grip; the wind is lighter and more easterly than predicted. But that has not stopped the birds. To the west I spot a sharp-shinned hawk beating out its familiar pattern: flap-flap-glide, flap-flap-glide. A second sharpie follows a few minutes later. A period of quiet, then a merlin flies low down the center of the island. It swerves slightly to the east, as if spooked by the sight of me, but still passes close enough that I can see the nostrils in its fierce, hooked beak.

The wind has sent these birds my way — or, rather, a combination of wind and oncoming weather has pushed them here. The flow of air between fronts is only one of several factors that determine the timing and course of animal migrations, but it is among the most important. This autumn show will not be repeated in spring. The regional winds will blow mostly from the south then, pushing the birds north in a broad sweep. Instead, the falcons and accipiters I have seen this morning will likely fly the Mississippi Valley to their breeding grounds, to settle in until the weather cools and October northers again propel them through the Outer Banks skies.

I like to think of the air that envelops and sustains us as being woven with shifting boulevards and tendril streets. I like to imagine an atmosphere of routes that remain invisible and largely ignored by people, but on which a host of creatures depend for transportation, for food, for signals about where to hunt or to hide from the hunter, where to seek shelter, where to woo and be wooed. These notions of mine are in no way farfetched.

To animals, especially the birds and flying insects, wind is a chariot to be ridden, a compass to be read. It is a source of knowledge, not only about present conditions but about weather soon to come. Nowhere is this more apparent than on a barrier island, where between the vast, unbroken horizons a careful observer can track the movements of meteorological systems. Weather fronts take on an unusually tangible quality here. You can mark their passage in processions of clouds; you can anticipate their tread in the disconcerting calm that precedes a radical shift in wind. And if you have studied the landscape, you can foresee their leverage on animals.

Last winter I spent a mid-January day in a duck blind with Vic Berg, a waterfowl guide who has hunted and fished on the Outer Banks for thirty years. Slim and sandy-haired, Berg keeps up an entertaining banter in the blind, masking the intense concentration with which he watches his surroundings. He is known as one of the best hunters on the islands.

The morning I chose to accompany him was one of the coldest of the year, with temperatures in the forties but with a leaden sky and a lashing northwest wind. We left solid ground at 5:15 and motored through dark to a marshy slice of land just inside Oregon Inlet. Directly north the pale yellow beacon of the Bodie Island lighthouse passed lazily over the landscape. "Good hunting weather," Berg said heartily as a weak dawn broke and we settled into an open-air box behind a sheaf of wax myrtle. "This is a real consistent spot. It's like there's a little funnel of wind that carries the ducks low past here on their way to feed in those ponds by the lighthouse."

A chevron of swans flew overhead, brilliant white against the clouds. A jagged file of cormorants made their way north, no more than a few wing lengths above the water. In the marsh behind us, a variety of rails might be calling. We could not say for sure. We could not hear them for the wind.

I asked Berg why windy days consistently make for better hunting.

"Because the wind limits the places the birds can sit on the water," he said. "You get a calm day and they'll spread themselves all over the planet. But a good breeze will eliminate three-quarters of their habitat. It stirs up whitecaps. Puddle ducks won't sit on whitewater at all; they want to find a little protected nook like that." He motioned to the shallows in front of the blind, where he had placed eighty handmade decoys. "There's a sandbar out there that knocks the waves down.

"Diving ducks will sit on whitecaps, but even they don't want to be on big rolling waves. They don't like getting slapped in the face. Then there's the way the wind affects the water levels. If it blows water out of an area so the grasses are exposed, it's like inviting the ducks to their own cafeteria."

I shivered as a gust blew my hood off. Turning to scan the sky in back of him, Berg seemed oblivious to the cold.

"Ducks ride the wind," he continued. "You can watch them do it. Sometimes you'll see a flock of birds going along one corridor and dropping down, and then another will come right along the same corridor. When that happens, you go set yourself down right below it, because there's a wind shear or something there. It's like the birds got together for a conference that morning and all agreed to fly the same route."

A quarter mile west a cluster of pintails and widgeons skated quickly by, ignoring the calls Berg made with one of three whistles he wore slung around his neck. A pair of mergansers passed low in front of the blind, but Berg held his fire. "They're not very tasty," he said.

I told Berg of a conversation I had held a few days earlier with a fisherman friend, who described how fish in both the ocean and the more protected, shallow bays of the coast react to subtle shifts of breeze. On the first navigable day after a summer northeaster, my friend said, offshore charter boats always seem to catch tuna, as if the wind forced the fish to school up tightly and move closer to shore. A strong northwest wind in autumn, on the other hand, draws red drum to the west bank of Hatteras Island.

"That's true," Berg broke in. "The water gets all churned up and the drum know there'll be lots of crabs in it. It's the opposite with flounder fishing — if you want to catch flounder you've got to look for clear water. In summer when it's blowing southwest, you can forget fishing in here. They'll be catching flounder in Manns Harbor, because that's where the water will be clear. Yep." He tipped his hunting cap back and pushed up his sunglasses with his thumb. "You can have all the best equipment, but there's no substitute for being where the birds, or the fish, naturally want to be. Which is where local knowledge comes in. Which is why you need a local guide."

A self-serving statement, perhaps, but it carried the ring of truth. I thought of some of my own local knowledge — that warblers of all stripes, migrating south in early autumn, tend to be funneled into a few squat hedgerows of wax myrtle on the Pea Island refuge. I looked out to the seething waters of the inlet and thought of the fish making their way invisibly through currents below. Moving water functions very much like moving air, not only in terms of physics, but in the stress it exerts on animals. In that barreling wind, on a morning of steely water and sky, the islands of the Outer Banks seemed like an afterthought in a fluid world, the dark, empty space between two parted lips. We're anomalies, I suddenly thought. We landlubbers are odd, clumsy beings suspended between two firmaments lush with flying, floating beasts.

SINCE WIND is born from a series of roughly predictable events, biologists believe some animals may use prevailing breezes to navigate through their worlds. This tendency may be particularly strong in regions of constant directional wind, such as the trade belts. In experiments with loggerhead turtles, researchers in Florida have found that hatchlings released into the Atlantic swim directly into waves generated by easterlies and so move swiftly offshore into the Gulf Stream. Some species of African desert ants appear to rely most heavily on the position of the sun to retrace their steps to their nests, but on overcast days they home to the wind.

Of the many ways animals react to wind, none is so well studied or so perplexing as the phenomenon of avian migration. Scientists know that prevailing wind patterns have helped shape major migratory flyways worldwide and that extreme winds can carry small species thousands of miles outside their normal range. After strong westerlies, North American butterflies, insects, and songbirds frequently show up in numbers on the British Isles. Beyond that, however, knowledge of the way birds move in wind is surprisingly scant.

Before the 1950s, biologists who studied bird migrations relied solely on seasonal census counts and the retrieval of bands, usually from dead birds. Much of the information known about migration routes today has been culled from the immense data banks of the U.S. Fish and Wildlife Service's Bird Banding Laboratory in Laurel, Maryland. But banding studies yield information only about the gross movements of bird populations, not about the singular environmental cues that might spur them on their way and guide them. Ground censuses, even when conducted by consummate bird watchers, tend to overestimate the numbers of birds moving through at low levels and miss completely those that travel at higher altitudes. (Early researchers speculated that headwinds stimulated birds to migrate, because so many birds fill the skies when southerly winds blow during fall migrations. It did not occur to them that migrants might simply fly closer to the ground during headwinds.) Thanks to radar, investigators now know that birds may fly at altitudes of several miles if the winds are favorable. John Richardson, an environmental consultant based in Toronto, has tracked shorebirds migrating to the West Indies in the rarefied atmosphere of 20,000 feet.

But radar studies have their own inherent limitations. For one, observers generally cannot identify individual species unless the birds happen to pass in front of the moon or through a ceilometer, a beacon of light pointed straight up. For another, radar tends to miss many low-flying birds and so skews counts in precisely the opposite direction of ground observations. "One of the biggest problems with this whole field of research is that it's nearly impossible to design good experiments," Richardson says. "In migration studies we're dealing with a huge number of species, with different age groups — experienced migrants versus inexperienced ones — and with routes of different lengths. And the birds may act differently on each leg of their trip."

In general birds will wait to embark on a migration until they can fly with a tailwind and minimize the energy they must expend. This is especially true of small species, which can easily be drifted off course by wind, and of those that must travel over large bodies of water or great deserts. Since birds can sense minute barometric changes through the pressure in their ears, they have a natural device for tracking weather systems and perhaps for figuring the direction of winds. Indeed, they appear to be so sensitive to such meteorological cues as temperature and humidity levels, cloud cover, and barometric fluctuations that some people believe they may be able to accurately predict the approach of a front and to use the corresponding wind shift as a navigation guide. As they travel, patches of turbulence and rolling buoyancy waves tend to bunch them, as one biologist noted, like flotsam on the sea.

Rather than fight a headwind, a small bird on a long flight may settle down to wait for a weather change — unless it is passing over inhospitable country or water. Sidney Gauthreaux, Jr., an ornithologist at Clemson University who has conducted radar studies since the 1950s, has seen thousands of warblers killed by the onset of a late "blue norther" during the spring migration over the Gulf of Mexico. "Most warblers have an air speed of only about twenty knots [twenty-three miles an hour]," he says, "and when they meet headwinds of thirty knots, they lose ground." The birds beat their way northward as close as they dare to the water, where the wind speed is lowest. They land as quickly as possible, perching in miserable huddles on light towers, ship rails, and drilling platforms. "I've seen oil rigs covered with dead and exhausted birds after one of those storms," Gauthreaux says.

In the 1960s Gauthreaux began an intensive radar comparison of prevailing winds and the flight of songbirds through parts of the Southeast. In surveys at Lake Charles, Louisiana, he found that during the peak of spring migration surface winds varied but blew predominantly from the southeast. At 3,000 feet, however, the winds were steadily from the south and blew at an average of eighteen miles an hour. The upper-level winds were clearly more suitable for migration, and Gauthreaux found that many species sought them out.

The arrival of a strong cold front in spring grounded the majority of migrants. But when birds met a weak cold front they simply climbed to higher altitudes and continued flying above the wedge of northerly air. Larger, faster birds such as waterfowl and shorebirds were less hindered by headwinds than were songbirds. Gauthreaux also observed that soaring birds like hawks and vultures traveled in the middle of the day when thermals were strongest. Slow fliers migrated at night, when they did not have to buck streams of upward-flowing air. He came away from the study convinced that birds seek out not only the most hospitable environments on land but the most favorable currents in the ocean of air. "Scientists have been naive," he says, "in thinking that all species use wind in the same ways."

A number of biologists are also rethinking conventional theories about drift, which occurs when strong winds force migrants off course. If birds always strive to fly the shortest routes possible (and until quite recently it was assumed they did), then through much of their journeys they might encounter powerful crosswinds and, occasionally, headwinds. The strongest, fastest fliers can push on; songbirds cannot. Often they simply take refuge on the ground and wait out contrary breezes. This tendency is what makes spring and fall bird watching so exciting in certain regions. But migration is a delicate business. If a bird's timing is off even slightly, it may get trapped by inclement weather or may arrive in its breeding territory too late to raise young. Many researchers now believe that birds are somehow equipped to wend their way through air currents that might otherwise carry them far afield.

In 1979 a Swedish scientist named Thomas Alerstam proposed a novel theory about wind drift that only began to win favor in the 1990s. Alerstam noted that being carried off course would pose significant problems only if birds steadfastly tried to fly as straight as the proverbial crow. But what if certain species followed elliptical flight paths that tracked prevailing regional winds? A bay-breasted warbler that had bred in Ontario, for example, might wearily buck crosswinds on a rhumb-line flight over the Midwest to reach its wintering territory in Central America. Or it might coast with seasonal northwesterlies to the eastern seaboard, as if it were riding a great chute. Alerstam developed several models to show how birds might make the best use of wind. Early in a flight they might glide at high altitudes, even if this meant they would drift far to the west or east. Closer to their destinations they could drop down and ride lighter, more variable surface winds. On its way north in spring, my hypothetical warbler would make use of prevailing southerlies to fly an arching route west of the Appalachian Mountains. Migration tracks might vary from year to year, depending on weather patterns. Such a strategy would not only be easier, Alerstam noted, it might be faster. But as John Richardson notes, "There are still no theories in this field that can be considered irrefutable. With all the work we've already done, with all the tools we have at hand, we still just don't know enough."

EVEN THE MOST favorable wind will not lead a bird directly to its goal, but only within several hundred miles at best. To find its way more precisely, it must depend on a range of navigational indicators. Teasing out the techniques by which birds navigate has proved to be one of the most difficult problems in ornithology, again because it is impossible to design experiments that mimic all the environmental stimuli a bird encounters on an extended flight. However, a few researchers have captured migrants and tested their abilities to orient themselves in a variety of settings.

In the early 1950s Gustav Kramer, a German scientist, found that migrating starlings trapped and held in round cages showed a strong tendency to face in the direction of their destination. By manipulating lights around the cages, Kramer showed that certain bird species depend heavily on the position of the sun to set their courses. Later studies determined that birds may also use polarized light, the positions of stars, and the direction of prevailing winds. Experiments conducted with that master of navigation, the homing pigeon, found that some birds may construct mental maps using local wind-borne smells and sounds.

Other work has demonstrated that pigeons respond to infrasounds, which are below the human range of hearing. The atmosphere is a constant thrum of wind-borne noise — the crash of ocean surf two hundred miles distant, the rush of air over a chain of mountains, the vibration of sand on desert dunes. Some biologists believe such sounds may be among the acoustic guideposts by which migrating birds select their routes. But as Gauthreaux notes, "Manipulating a cue and having birds respond to it is not the same as proving that the birds actually use it in navigation. That's making a rather quantum leap."

What has emerged from decades of scientific probing into the interactions of birds and wind is a literature of patchy and often contradictory findings. Experiments conducted by Gauthreaux and a colleague in the early 1970s showed that songbirds passing at night through the southeastern United States invariably flew with the wind, even when this behavior caused them to fly back toward the north in autumn and back toward the south in spring. Later studies found that songbirds in Canada tended to buck winds that would have carried them in the wrong direction for the season. Why might the same species show such utterly contrary behavior on different legs of their migration? Gauthreaux notes that seasonal winds in the Southeast are more reliable for migration than those in Canada. "When birds reverse-migrate in the Southeast, they do so for only short distances," he says. "They very quickly turn around and head in the right direction. I've long thought that migrants may use different strategies as they get closer to their goals."

Given that birds spend their entire lives on the move, alert to each small shift of weather, it is not surprising that we should find aspects of their behavior puzzling. Their unfailing ability to make their way to far-flung forests and velds remains remarkable, one of the grand mysteries of life.

LATE ONE APRIL afternoon, when columns of sun-heated air have set misty shadows dancing on my living room floor, I take some work outside to the screened porch. The air is fresh with spring; the dogwoods are in full bloom. My resolve to work slips away like water through a sieve. I make a stab at reading but find my gaze wandering to the play of sunlight through the trees, the cat lolling in the yard, the skittering flight of a Carolina wren. For five years a wren couple (perhaps this same bird and its mate) has nested under the eaves of the house. Unabashedly idling now, I watch the wren pick among pine straw. Then abruptly I focus on a glinting thread that drifts a few feet away in the heated air.

A strand of spiderweb — no, two strands, a foot apart — wave from the azalea in front of the porch.

I know suddenly what these are. I remember the baby spiders explaining their actions near the end of E. B. White's classic children's book *Charlotte's Web:*

> "We're leaving here on the warm updraft. This is our moment for setting forth. We are aeronauts, and we are going out into the world to make webs for ourselves."
>
> "But *where?*" asked Wilbur.
>
> "Wherever the wind takes us. High, low. Near, far. East, west. North, south. We take to the breeze, we go as we please."

Charlotte and her daughters were identified only as gray spiders, "about the size of a gumdrop" when grown. A few days after hatching they let loose whole balloons of web and drifted away, leaving Wilbur, the pig, distraught below. The spiderlings in my bushes (if indeed any are there) have each released a single thread. I go outside to see what anchors them, only to find them vanished. I run my hand through the air where I gauge them to be and grasp nothing.

But when I climb back onto the porch they reappear, waving lazily, silver gossamer in a tongue of light.

Maybe these are *Epeira* spiders, the stout-legged, cosmopolitan creatures studied by Jean-Henri Fabre in France early in this century. In his book *The Life of the Spider,* Fabre wrote of watching the cross spider, *Epeira diadema,* play out a single thread, roll it up into a little package, climb to a high spot, and wait for the wind to unfurl it. By heating the air in his laboratory with a chafing dish, Fabre created thermals that sent the young spiders sailing up to a twelve-foot ceiling. Then he opened a window and set them free on the wind.

It is marvelous to imagine the life that might be floating in any cubic yard of air, life that owes its continued existence to the drafts stirred up by sunlight. Spiders, butterflies, aphids, beetles: all ride currents of air from their places of hatching to suitable niches nearby or far. Some cross whole watersheds or come to rest on remote ocean islands. They embark not on brisk winds but on lazy upwellings that push like bubbling creeks into the air up to two hundred feet and sometimes higher.

Thirty miles southwest of my home, on the swampy edge of the North Carolina mainland, the U.S. Air Force operates a range where pilots practice dropping bombs on targets of fetterbush and fern. In the spring of 1994, air force officials hired a British falconer and radar expert named Adam Kelly to monitor the movements of the area's bird population. The purpose of the study was to reduce the odds of collision between aircraft and birds, especially the larger species like turkey vultures and trumpeter swans, which could conceivably down a plane.

Kelly set up camp in a trailer next to a bunkerlike building that serves as base headquarters. Beside the trailer was a narrow blackwater canal and two radar towers with revolving cones. Bleak, tick-infested land lay in all directions. "People told me it was as ugly as a lunar landscape out here," he says. "The first afternoon I drove out, I saw a sharp-shinned hawk passing over about every forty seconds. Then I drove back to Manns Harbor and saw the sun setting over

Croatan Sound. I thought, I could stay here quite a long time. This is paradise."

One morning in late March, Kelly noticed hundreds of small green blips on the screen of one of the radar units. He stepped outside to scan the sky but could see nothing. Puzzled, he lay down on his back next to a conical scan tower and studied the air through binoculars. "I didn't see many birds, maybe one every couple of minutes," he says. "I was looking really hard; my eyes were watering. Then it occurred to me that I was seeing an awful lot of butterflies." There were thousands, flying in dense clouds between 750 and 1,500 feet, in light southerly winds. Without the radar he never would have noticed them. "I think, in general, large-scale insect movements are completely unseen by people," he says. "For one thing, it's hard to tell from the ground exactly what's going on up in the air. But also insects are small, and we're not trained to pay attention to them. In fact, we're effectively trained to ignore them."

Kelly adjusted the gain on his radar units so they would pick up only targets the size of birds. But on a sultry afternoon two months later, another unit was blacked out by a sudden insect swarm. "A farmer just south of here had disked a field that morning," he says, "and the sun shining on the dark soil set up a bunch of convection currents. You could see the insect swarm on the screens, spreading up from Hyde County like a fungus.

"We drove past the field, and the truck was literally covered with bugs."

No other class of animals lives so starkly at the mercy of wind as the insects. In its benign, convective form, wind plucks up bugs by the millions, the trillions, every day and scatters them widely. (This is not to say they are dispersed at random. Biologists have often seen insects choose a precise moment to set out — say, just before the onset of a light rainstorm with west-bearing winds — presumably so that they are most likely to arrive in protected, food-rich habitat.) Once aloft they depend on horizontal currents to ferry them onward. Often they drift all night. In the upper strata they encounter waves of turbulence and layers of different densities that sort and compress them into compact swarms. They may drift as a unit through honeycombs of thermals and land lightly in a rogue's paradise — a flower garden, a newly cultivated field of vegetables. Or they may be accosted by a freshening gale, tossed out to sea, and slammed rudely into the water.

Many insects are wingless, at least in their larval forms, and must depend entirely on wind for transit. Those that can fly have neither the mass nor the strength to fight headwinds. So they ride the breeze, perhaps passively, perhaps with some precise collective plan. In certain species, individuals bunch together like flocks of birds, turning as one organism. In studies conducted during the 1950s, aphids carried by west winds from the irrigated oasis of southern California into the arid basin of the Colorado desert showed an uncanny knack for finding potted alfalfa plants left out for them. Some investigators speculate that spiders and other balloon-riding creatures may control their altitude by trimming the lengths of silk that lift them. At present, however, scientists do not know whether insect wayfarers have anything but the crudest ability to navigate from one habitat to another.

Until the 1920s researchers were unaware that insects can survive well up into the troposphere. It had been known for a century that the smallest insects could be easily transported by wind. In 1827 and again in 1924 British Arctic explorers discovered live spruce aphids and syrphid flies on the Norwegian island of Spitsbergen. Apparently they had been carried by wind from Russia, 800 miles away. Monarch butterflies had been observed 500 miles from land over the South Pacific, and once a swarm of migratory locusts was seen over the North Atlantic, 1,500 miles from land. With the spread of European gypsy moth larvae through the eastern United States during the early 1920s, research into the dispersal of insects took on special urgency.

In 1926 Perry Glick, an entomologist in Louisiana, made a series of flights in which he sampled the fauna of the air with traps made of wire screen mounted on the wings of a small biplane. Few scientists had tried to examine life in the upper atmosphere, and Glick's study is regarded as a pioneering work. At 200 feet he found a great number and diversity of insects. He caught by far the most when wind velocities were between five and eight miles an hour. Lesser winds seemed to discourage insect flight; greater winds made it nearly impossible.

Above 300 feet the insect populations rapidly thinned. Nevertheless, Glick found a few spiders and flies as high as 5,000 feet. He never devised an accurate means of measuring the flow of convection currents in the upper atmosphere. But on days when the plane encountered turbulent air, he caught many more bugs at 1,000 feet, as if strong thermals had tossed them to unusual heights. Over five years Glick trapped 700 identifiable species. In the end he calculated that on any midsummer day the air above each square mile of Louisiana contained more than 14 million insects.

Many insects travel at night, presumably to avoid the upper-air turbulence caused by convection. Most individuals migrate no more than five miles over the course of their lives. Only a few species embark on lengthy seasonal movements, and then largely because of their proximity to favorable avenues of wind. Monarch butterflies, migrating from the eastern United States to their overwintering grounds in Mexico, follow routes that parallel the major flyways used by birds. In the North American interior and in northern China and Japan (all regions where winters tend to be particularly severe), certain butterfly species ride jets of warm air north with the coming of spring.

The best-documented relationship between migrating insects and wind involves the locusts of the African deserts, which move in great swarms with seasonal shifts of the Intertropical Convergence Zone. In biblical Egypt a plague of locusts was delivered by east winds; and indeed wind is one of the key factors (although far from the only one) that governs the occasional explosion of locust populations and the wanderings of these voracious crop pests throughout the world.

The still, moist air of the convergence zone floats like a shadow over the earth's midsection, varying in width between 350 and 550 miles, changing position with the annual tilting of the earth. Over land its movements are closely accompanied by hordes of insects. In Africa, locusts abandon regions that are growing parched and travel in great, rolling swarms, following moist air for hundreds of miles. Other locust species invade wet portions of West Africa just north of the convergence zone but avoid its southerly reaches because the heavy rains there drown colonies and promote the growth of lethal fungi. Such large-scale movements, pegged clearly to seasonal wind patterns, may serve as a paradigm for less obvious ecological relationships. Perhaps insects all over the world respond to subtle weather changes — temperature and humidity fluctuations or wind shifts too small to be much noticed by people — by setting themselves adrift en masse.

SINCE THE 1970S the wind's role in the transportation of insects has become a major focus of efforts to protect forests and crops from such pests as mites, scale, caterpillars, and gypsy moths. The most promising control methods involve the use of pheromones, those invisible, airborne love potions on which the bug world seems to depend as heavily as humans rely on flirtatious winks and smiles.

Shortly after the turn of the twentieth century, biologists discovered that insects widely use what were then called sex attractants to lure in mates. Despite the significance of the discovery, subsequent research proceeded slowly. In the 1950s, investigators at last succeeded in isolating bombykol, the primary sex pheromone emitted by the silkworm moth.

The field expanded dramatically during the 1970s and early 1980s. But pheromone research demands particularly exacting and patient technique. The subjects are small and delicate, and even in the best conditions they release minute amounts of the chemicals under examination. In addition, it is difficult to prove that a certain substance elicits a presumed response. Working in wind tunnels, biologists can manipulate one variable at a time. They have been able to calibrate the exact conditions under which different species are most likely to play out plumes of odoriferous bait. But they cannot watch how scents disperse through field and forest or from what distances suitors fly upwind to find the source.

Nevertheless, from nearly a century of sporadic research scientists have reached a few salient conclusions. First, to create a plume an insect must fan its scent, either by flapping its wings or releasing the scent into the wind, so it requires less energy to stroke pheromone-producing glands during periods of gentle breeze. But if necessary the insects will begin "calling," as biologists say, in dead calm. If a heavy wind springs up, the release of pheromones stops as the insects seek cover. Second, each species sends out scent signals on select "channels" to which its own kind is listening. That is, it emits particular compounds of odors in distinct rhythms at specific times of day and only in certain seasons. This phenomenon has been particularly well documented in butterflies and moths.

To follow a pheromone trail, an insect does not set a straight bearing but engages in a sophisticated pattern of zigzag flight that apparently takes it in and out of the odor plume. A scent is like any other substance released into the air; a gust of wind may stretch it thin or send it curling backward. By casting back and forth, insects apparently can gauge how an odor has been pulled by the wind and better follow it to its source. Recent wind tunnel experiments by Thomas Baker and Neil Vickers of Iowa State University have found that when male tobacco budworms encounter strands of odor-laden air, they will fly hard into the wind. But when they are surrounded by continuous clouds of pheromone scent, they seem to become disoriented, and their flight toward the pheromone source stops. It is as if both the pulses of odor and the pockets of clean air between them serve as directional signals. The faster the male budworms encounter whiffs of pheromone, the more quickly they fly to its source.

A number of pheromones have been manufactured synthetically and eventually may be widely marketed as natural pesticides. One method calls for permeating a breeding habitat with a single scent, effectively jamming the airways so pests cannot find mates. Pheromones have also been used in traps to monitor the spread of insect populations and, with less success, as bait in population control experiments. Following a plume upwind, the would-be perpetrators become trapped in cages or on sticky boards, the quintessential prisoners of desire.

WIND IS God's agent of transport, a living broom that sweeps flying animals into great swirls and piles. It carries them over rivers and seas and on rising columns up mountainsides. At times it sets them down in barren, high-altitude nooks where only the hardiest organisms can survive.

Lawrence Swan is professor emeritus of biology at San Francisco State University. Early in an introductory conversation he tells me he was "born and raised in the Himalaya" (which he pronounces distinctly as "Hi-MAL-ya — it means the abode of snow in Sanskrit"). He has been smitten with mountains all his life. On their honeymoon in 1946 he and his bride explored Citlaltepetl, an 18,700-foot peak in Mexico. "We got separated and I left her on top by mistake. She hasn't let me forget it since," he says.

In 1954 Swan was part of the first American expedition to the Himalayas after World War II. He had a particular purpose for making the trip. During the honeymoon on Citlaltepetl he had noticed some lizards feeding at unusually high altitudes, and he wanted to see if he could find a similar species in India. Also, thirty years earlier a British expedition had discovered a curious jumping spider on Mount Everest at 22,000 feet. All spiders are predacious. What could they have been feeding on, Swan wondered, at a height of four miles? As he and his companions made their way tediously up Mount Makalu, he paused to root through a six-foot pile of rocks. "It's a sort of biological dictum that animals can't survive in a region without green plants," he says. "So I started digging around a bit and found some little Collembolas — some springtails. And my gosh if there wasn't a little sheen of something on the rock. I took a piece of paper and scraped it." The sheen turned out to be pollen. In the crevices between rocks Swan found thin accumulations of plant fragments, seeds, spores, and bits of insects. Food carried to the mountain by wind.

Six years later Swan returned to the Himalayas for an extended trip. Over six months he walked a thousand miles through the mountains and compiled one of the most extensive collections of Nepalese animals and plants in the world.

He also confirmed his suspicions that a previously unknown ecological system, one supported entirely by wind, exists in the most baked and oxygen-starved landscape on earth. In subsequent work on Citlaltepetl, Swan described the terrestrial, aquatic, and nival — that is, snow-locked — components of what he called the aeolian biome. "I named it, of course, for the Greek god of the wind."

The aeolian zone begins a bit above the tundra zone and consists of a patchwork of isolated ecological niches that are created where currents of wind, having lost speed and strength, drop a rain of algae and insects. At times the snowfields of the aeolian zone turn green or pink from the accumulation of wind-dropped microscopic plants.

The algae and insect carcasses attract tiny but sturdy scavengers — salticid spiders, daddy longlegs, jumping bristletails, annelid worms that can survive encased within ice. These in turn feed lizards, salamanders, rattlesnakes, and a few species of birds. Snowmelt forms pools inhabited by delicate springtails and fairy shrimp. Where water pours in muddy torrents from the mouths of glaciers, stoneflies congregate in dense clouds, feeding on wind-borne debris that may have been trapped in the ice for hundreds of years.

With the intense solar radiation and the thinning atmosphere, life at such heights must withstand extreme swings in temperature. Bacteria collected for Swan at 27,600 feet by a 1963 expedition to Mount Everest showed adaptations to freezing and thawing each day. One form was so unlike any other on the earth that it defied classification. "If there could be any wind-blown organic nutrients on Mars," Swan wrote, "this or a similar organism would be the best candidate for survival on that planet."

In 1992 Swan published a paper in the journal *BioScience* summarizing his career's work. In the conclusion he suggested that wind-dependent life may be common throughout the world. In particular he mentioned the interior of Antarctica, where wind-borne nutrients could conceivably support an assemblage of animals too small to be discerned by casual observations or from photographs. By recognizing that similar aeolian ecosystems have evolved in the coldest, most far-flung regions of earth, scientists could incorporate Antarctica "into the normal domain of biogeography," Swan wrote. "The word *aeolian* could also replace such mapmakers' terms as *barren, glaciated,* and *snow or ice* used to mark the lands beyond the tundra and that imply a misleading sterility."

In closing, Swan asked a provocative question. What if the first terrestrial organisms were not plants, as is commonly believed, but animals that fed on nutrients carried to them from the sea by wind? He noted that even now organic-rich compounds from evaporated sea foam can be found in snow on some of the highest peaks in the world. He described a bristletail he had discovered living on the Barun Glacier in the Himalayas "among the tumbling rocks between the moving ice crevasses in an environment that would challenge comparison with the niche of any other animal. It survived on windborne debris, a frontier creature of the aeolian region." Might it be related to the earliest bristletails of the Devonian period? Might not creatures that evolved 390 million years ago also have lived solely by grace of wind?

The paper elicited dozens of requests for reprints, but to Swan's disappointment no scientific response. "You can show someone an aeolian system, and at first they won't believe you," he says. "They don't think anything's really there. It takes a while for the concept to be absorbed by the mind — for even astute scientists to realize just how many insects and animals survive on what is brought to them by the wind."

IT IS ANOTHER April afternoon only a few days after I watched the spider silk wafting from the porch. I am thinking not about animal migration or aeolian biomes but about a more immediate problem: how to stay warm in a biting breeze. I walk stubbornly north along a favorite beach, dead into a torrent of air. The surf has lost its midwinter blue; a new bloom of plankton has turned it a bright, deceivingly warm-looking jade. I move stiffly, my hand gathering the hood of my coat just below my chin to keep it tight against my scalp. I once read that the forehead acts as a comfort gauge for the whole body: if your forehead is cold, the rest of you is freezing. Just now I believe it.

I find the signature of wind everywhere I look. Near the water the sand surface is crusty. The smallest grains have been pitched west onto a line of dunes, where they tumble on the edges of gusts. Tiny, knife-shaped ridges jut from behind strewn shells. I give up on walking north and turn back south. A few sanderlings hop along the tideline, searching for food. Ahead of me I notice more, a flock of perhaps three dozen. I can't see them very clearly for blowing sand. They stand in a tight group, faces to the wind, appearing and disappearing in clouds. I think of a picture I once saw of a telephone pole in the desert, its base eaten away by sand. If windblown grains can devour treated wood, what damage can they inflict on a shorebird?

The answer, I find later, is very little at all, as long as the bird points itself like a weathervane into the breeze. Otherwise sand will lodge between its ruffling feathers and reduce its ability to stay warm. In heavy wind sandpipers may seek shelter behind small rises, but even then they stand together facing the adversary, their eyes covered with retractable membranes. Being close to the ground, they do not feel the full force of the wind. Nonetheless, they live at a level where high turbulence sends sand in unpredictable swirls.

Every wild creature has evolved some means for protecting itself from the chilling effect of wind or, in hot climates, for seeking wind out. Bobwhites on the Great Plains take refuge behind tufts of grass during winter storms. Bald eagles in the Pacific Northwest have been known to fly for miles from the rivers where they feed to overnight roosts in coniferous forests, which offer more solid shelter than deciduous woods. In the scorching Namib Desert, antelope and beetles climb to high spots in late afternoon to refresh themselves in sea winds stirred up over the western Atlantic, and nest-sitting ostriches raise themselves a few inches so flowing air will cool them and perhaps their eggs.

Wind reliably scours snow from certain slopes in the Rocky

Mountains and provides winter grazing for bighorn sheep and goats. It massages the animals' heavy fur and keeps them from becoming overheated in the intense radiation of sunny afternoons. But it is always a treacherous companion. The scientific literature is full of stories of animals frozen in their tracks by the sudden onset of brutal gales, of hurricanes tossing small birds by the hundreds into buildings, of gusts dashing sea ducks against rocky cliffs.

To me the most intriguing accounts are those in which animals turn the presence of wind to their advantage.

Along the sandy roads and dune swales of the southern Outer Banks lives a curious arachnid known as the turret spider. Surrounding the entrance to its two-foot-deep burrow is a thumbnail-high, silk-lined mound. Turret spiders have long been a subject of interest to Steven Vogel, a zoologist at Duke University. Years ago, on a trip to the spacious beaches of the North Carolina barrier islands, Vogel began poking around their burrows. "Why do they build that turret?" he asked himself. "Flood protection? It's only a few centimeters high. A trap? Probably not; it seems too obvious. I think it's drawing air."

Back at Duke, Vogel constructed a tiny anemometer that could be lowered an inch or so inside a turret spider's burrow. On his next trip to the beach he found that even a light breeze passing over the burrow entrance caused a draft to flow upward through the mouth. "Sand is a loose soil," he says, "and its grains are surrounded by moist air. The spider has designed a simple, wind-driven system that causes air to flow from between the sand grains into its burrow and out the top." In the process it has tapped into a source of moisture in a desertlike habitat.

It occurred to Vogel that a variety of burrowing animals might purposely build their living quarters to create currents of air. One summer he contacted a biologist whose family had a farm in Kansas and asked for his help on a study of the living quarters of black-tailed prairie dogs.

To stand in the midst of a thriving prairie dog town is to feel ogled by a hundred pairs of hidden eyes. The dry, cobbled soil is studded with holes that lead downward, to depths of perhaps ten feet. Most burrows consist of a single passageway that joins two openings, one in a thigh-high earthen mound and the other in an unobtrusive hole, often hidden beneath a clump of sage or grass. Guards click out staccato warnings from the tops of the mounds, which biologists long assumed served mainly as flood guards and vantage points.

"I started wondering how animals that spend so much time underground could meet their requirements for oxygen," Vogel says. Using measurements provided by the Kansas biologist, Vogel and a colleague built a model burrow inside a Plexiglas chamber equipped with a fan — a simple wind tunnel. "The model was one-tenth the size of a real burrow," he says. "Then we increased the wind speed ten times over what you'd normally find on a quiet day on the prairie." The greater wind speed compensated for the reduced scale of the underground passage.

The results were surprising, even to Vogel. Like the spider turret, the mounded entrance created an upward draft whenever wind passed over it. As a result, air was pulled into the smaller, hidden entrance. The shape of both openings, as well as the angle at which the passageway descended, seemed designed to take optimum advantage of this current. Even a soft breeze of one mile an hour was sufficient to completely change the air in the burrow every ten minutes.

In 1978 Vogel published an article in *Scientific American* in which he described nearly a dozen ways that animals use moving air and water to cool or feed themselves. In addition to turret spiders and prairie dogs, he wrote of African termites that stud their ten-foot-high mounds with ventilation holes that funnel cool air to interior brood chambers, of sponges that tailor their shapes to take the best advantage of currents filled with nutrients, of a limpet with a "keyhole" opening that induces a flow of water under the lip of its shell and out the top. "Both air and water are viscous fluids," Vogel says, "and they react similarly to the laws of physics.

"The fascinating thing to me about all this is that you've got animals using technology in ways very similar to humans," he adds.

"In the Australian Outback opal miners build underground houses designed to take advantage of airflow. Who does that sound like? In Iran there are houses with openings around their roofs that draw air through them for cooling. It's all achieved through the same principles."

ON MY WAY HOME from the beach this April afternoon I drive across a high, arching bridge. From the summit I can see the ocean behind me and, ahead, the mile-wide marsh that fringes Roanoke Island. I am not, it seems, the only one enjoying the view. On the east side of the bridge three gulls hang in midair, pointing into the wind. They are neither gaining ground nor searching the water for fish. They look for all the world as if they are having fun. And why not?

Why not steal a few moments to go hang gliding if you are a gull on a spring afternoon? Why not harness a bit of the wind's strength for pure fun? Wind comforts us and challenges us and feeds us, physically and spiritually. It helps power the engine of life. I drive home in a celebratory mood, my spirit buoyed high on waves of air.

FIVE

The Voices of Trees

ON THE NIGHT of August 24, 1992, the Florida coast just south of Miami was swept by a tempest more than a decade overdue. The eye of the storm known as Hurricane Andrew passed south of central Miami, sparing the old art deco hotels, the financial district, the most populous neighborhoods. The tightly circulating core, with winds clocked at more than 145 miles an hour, crossed southern Dade County, rolled over the towns of Florida City and Homestead, and spun through Everglades National Park. In terms of the damage wrought to natural landscapes, the track of the storm could hardly have been worse.

The following day a biologist named Joe Maguire and several colleagues climbed to a catwalk atop the Charles Deering mansion on Biscayne Bay to survey the ruined land. The Deering estate, built between 1918 and 1922, was a botanical treasure, with 378 acres of hardwood hammocks and slash pines rooted in ancient marine rock. The estate had felt Andrew's full power: the tidal surge had crested at seventeen feet, high enough to send waves crashing against the stone mansion's second-story windows. A towering pile of debris — broken branches, splintered lumber, even a couple of cars — had been tossed beyond the mangroves that fringed the shoreline into the hardwood hammock. What trees still stood were stripped of leaves; their naked, twisted forms looked wasted and otherworldly.

The only sign of life could be seen a bit to the west, where a stand of slash pines still held most of their long, elegant needles. "We all thought, Great, at least the pine rocklands are going to make it," Maguire said. "At least they've got a good chance."

But over the next nine months the pines began to sicken and die. As the mangroves sent up a bright covering of new leaves and the tropical hardwoods again budded, as vines threaded into groves once shaded by giant gumbo-limbo and wild tamarind trees, the pines dropped their needles and perished. They died beneath the assault of millions of insects, especially pine bark beetles and pine reproduction weevils.

Insect infestations are common after cataclysmic weather events. The pine rocklands had survived many such attacks. But this time the trees could not adequately defend themselves. Something about their very makeup had been altered; they could no longer produce enough pitch to fill the holes drilled by the beetles and weevils. And so a little more than a year after Andrew, a boneyard of snags towered over a snarl of new, largely exotic growth. Between the dead pines and in the ruins of the hardwood hammocks, the vine known as air potato sprouted, and the tall grass called Burma reed, and the shrubby Brazilian pepper tree.

In the years since Andrew, Maguire and other conservation biologists have found themselves facing a formidable adversary in a war that continues long after the human communities have been rebuilt. It is a war against invading armies of plants, all of them well equipped to prosper in the fragmented forests of south Florida. All of them hardier and more aggressive than native species, and most of them spread by wind.

I AM ON my hands and knees in Everglades muck, strange patterns dancing before my eyes. Dried roots, sprouting grasses, delicate forbs I have never seen before and cannot name; tiny leaves with rounded edges, with serrated edges, with wispy purple veins. A crusty, fibrous accumulation of microorganisms known as periphyton coats the earth like a layer of dried paste.

For several hours now I have been combing a saw-grass prairie in the Chekika area of Everglades National Park, searching for the seedlings from a nearby *Melaleuca* tree. An Australian species with papery bark and long, elegant leaves, *Melaleuca quinquenervia* is a beautiful ornamental, and astoundingly quick to grow. Since the 1950s it has become a major threat to the native communities of south Florida.

I have come to this field on a January day with three biology students from Rutgers University who are trying to get a fix on how quickly *Melaleuca* can invade new areas. The landscape here is spare, even barren, a plain of saw grass colored with subtle browns and greens. The sky is dull with clouds. In the midst of the field a lone *Melaleuca* tree stands surrounded by bright pink and yellow survey flags. These mark the transects we have already scoured for seedlings.

Sylvan Kaufman, a slim Ph.D. candidate with a long braid, examines a study plot and marks a zero on her data sheet. I crouch over a nearby plot next to the other students, Michelle Hughes and Andy Bersch. Our eyes move slowly across the ground as we search for the slender-leaved seedlings Kaufman has shown us. They are far from obvious. "We take life to a whole new level of small," Hughes jokes.

I move aside a tuft of saw grass, careful not to cut my hand on the razor leaves. Beside it are two thumb-high, bright green *Melaleuca* seedlings. They are hale and hearty; their foliage glimmers even in the dull gray light. "I've got two here," I say, and my companions groan.

In eastern Australia *Melaleuca* is a rather quiescent member of an ecosystem that has been largely destroyed by development. But in North America *Melaleuca* is disconcertingly prolific, even in the worst of times. Every mature tree produces *billions* of seeds, and it may hold them for many years until some form of stress — a fire, a hurricane, a drought — triggers their release. They are then dispersed by wind, and perhaps also by water.

In April, eight months ago, this field was badly burned when a trash fire got out of control. Roughly fourteen days later, the

Melaleuca tree released its rain of seeds. Now we are scouring transects leading from the tree due north, south, east, and west. The largest number of seedlings took root to the east, presumably blown there by a westerly breeze.

The transformation of saw-grass prairie to *Melaleuca* thicket is an amazing phenomenon. One or two pioneers take root in new ground, looking no more like a threat than, say, an old oak in a grassy field. Then suddenly a dozen more surround the pioneers, then several dozen more. That is how exotics spread: insidiously, with great speed, and with what strikes me as cunning. Biologists call it the guerrilla effect.

Our survey of the prairie near this one tree has taken almost twice as long as Kaufman had hoped. We pull up the flags and hike to another tree a quarter-mile east, stepping gingerly on the squarish gray stones that litter the prairie like rubble. The saw grass thins. Dreary gray periphyton coats the uneven ground. Between the stone outcroppings are depressions where soft butterscotch mud waits to suck at our boots should we fall in. My vision swims with the odd shapes and colors of the earth, and I stop; for a moment I am too dizzy to move. I close my eyes for a few seconds, open them, and push on.

When we reach the tree Kaufman hands each of us some survey flags. "We're going to do a grid pattern, five meters on each side of the tree," she says. We mark off a row of ten meter-square survey plots radiating east and west, and ten radiating north and south. We flag more squares, and more, until the tree stands at the center of a hundred plots. Silently we each take a row and begin to scan the measured bits of ground, one by one.

ESPECIALLY IN temperate climates, wind has left its mark on plants and trees since they first became established on land. The earliest trees ever to lift their limbs were probably primitive organisms pollinated only by wind. And to this day the gymnosperms, the naked-seeded plant species that include coniferous trees, depend on wind to spread their dusty sperm. Each April the loblolly pines that surround our house let loose clouds of green powder that coats our yard, our cars, our porches, and any tools or toys left outside. Grasses grow enlarged, feathery stigmas that are well adapted to catching airborne pollen, and they produce seeds that can be easily spread by wind. Wind is so important to many crops, including corn, wheat, and sorghum, that agronomists have learned to plant rows perpendicular to prevailing breezes and in specific patterns to maximize pollination and minimize soil loss from erosion.

From the arid plains of the West to the temperate forests of the Northeast, plants send forth their progeny by way of the wind. The stems of tumbleweeds are designed to break off in storms, allowing the plants to spread their seeds as they roll. Orchids emit dust-like seeds that are easily carried by breezes. Dandelions and milkweeds produce the silky calyxes beloved by children, and maples cast forth the winged samaras eloquently praised by Henry David Thoreau: "In all our maples . . . a beautiful thin sack is woven around the seed, with a handle to it such as the wind can take hold of, . . . and this it does as effectually as when seeds are sent by mail in a different kind of sack from the Patent Office. There is a Patent Office at the seat of government of the universe, whose managers are as much interested in the dispersion of seeds as any body at Washington can be, and their operations are infinitely more extensive and regular."

But for plants, as for all other life, wind wields a double-edged sword. In addition to spreading species, it has immense power to dry stems and leaves and to transport salty, leaf-burning ocean spray. On mountain peaks and barrier islands, shrubs and trees grow twisted and humped from the constant roil of wind over their forms.

I have come to the outskirts of Miami to examine a biological anomaly created by patterns of settlement. In effect humans have drawn a line down the middle of south Florida, reserving the west side, the Everglades, for nature and the east side for decorative gardens filled with showy tropical plants. Exotic botanical collections are as much a part of Miami's culture as the Spanish language and Cuban sandwiches; they help create a wonderful melding of

North American and Caribbean textures. But plant communities do not recognize such boundaries.

To understand how radically the landscape of south Florida has been changed since the 1950s, one must remember that the tip of the peninsula is an ancient reef known as the Miami Rock Ridge. In its virgin form the ridge looks oddly like cement that has been partially eaten away by acid. Blocky protrusions of pinnacle rock, as it is called, rise and fall unevenly across the flat terrain. Solution holes as narrow as a few inches or as wide and deep as twenty feet lie hidden by vegetation, making it hazardous to walk or drive in unfamiliar territory.

For many years the presence of the ridge restricted dense urban development to the sandy soil of northern Dade County, where today virtually no natural areas are left intact. The rocklands of the southern part of the county were considered too uneven and inhospitable for building. But in the 1950s the invention of a chipping machine called the rock plow gave developers a means of grinding pinnacle rock into a pebbly soil that could be evenly raked for agricultural fields. In areas slated for housing and commercial uses, the rock-plowed soil was cleared away and replaced with sand fill. The explosive development that followed reduced the indigenous pine rockland forests outside Everglades National Park to 4,400 acres, less than 3 percent of the pine forest cover of presettlement times. Most of the rockland forests that remain today lie within county-owned parks and private preserves. The same fate befell the hardwood hammocks: outside of protected areas, they were cut.

After Andrew's passage, imported vines such as air potato and wood rose began infiltrating more of the hardwood hammocks every day. Mature hammocks have a very tight canopy with an open, shady area beneath that prevents exotics from getting established. But the storm knocked down so many big trees that exotic vines could prosper even in the thickest groves.

The Everglades, unlike the Miami area, appeared to have escaped major damage from the hurricane, at least initially. Most of the slash pines within the park survived. And before the storm a crew had established a four-mile-wide *Melaleuca*-free buffer on the east side of the park. "There was a lot of worry right after Andrew," says plant ecologist Jean Marie Hartman of Rutgers, "especially about wind-dispersed exotics. But what people forgot was that the exotic trees took just as much of a beating as the native ones. They weren't in any condition to do much invading for a while." Nevertheless, many mature hammocks had been knocked down, and biologists feared that invading species could still gain a solid hold.

In 1993, when Hartman began an intensive study on how the hurricane might have spread four exotic species (*Melaleuca*, Australian pine, Brazilian pepper, and an imported marlberry) within the Everglades, she quickly realized that little was known about the way exotics move into new territory. She needed to begin with very basic tests. Conditions the previous year had been unusually wet, which some people thought might discourage germination of *Melaleuca* seeds. But one of her students soon managed to sprout *Melaleuca* in standing water, overturning the conventional wisdom that the species needed dry soil to germinate. Within the park itself the species grew in areas with knee-high muck. "The literature is wrong," she says. "This is an aquatic plant."

The research center where Hartman's Everglades field office is based is in a section known as the Hole in the Donut, a 10,000-acre rectangle of invasive exotic plants surrounded by naturally pristine hammocks, prairies, cypress domes, and pine forests. When Everglades National Park was formed in 1934, this area was left as agricultural land. After the park acquired the property in the late 1970s, many of the abandoned farm fields were invaded by Brazilian pepper, a shrubby, exotic, and very prolific tree related to poison ivy. Its seeds are dispersed by wind.

The road to the research center is edged on both sides by what seem like endless Brazilian pepper thickets. The short, weedy trees grow so tightly together that I cannot see more than a few feet into them. The species soon reinvades areas that are cut or bulldozed or burned. "The only way they've found to get the natural vegetation to come back is to take out all the loose agricultural rock until they've gotten down to the base again," Hartman says. She shows me a field, completely devoid of cover, where pinnacle rock is once again visible beneath a patchy film of dirt. In the distance are long windrows of uprooted Brazilian pepper trees. Dirty yellow road graders roll slowly across the field, pushing loose stones in front of them. "Even this isn't good enough to keep Brazilian pepper out," Hartman says. "The last step is to come in with some brush equipment, kind of like street sweepers, to clean off the dirt. There needs to be nothing left but rock."

It is a sobering spectacle. The restoration of the Hole in the Donut will cost as much as $44 million and is expected to take until 2012. How sad to think that so much money and time must be spent to bring an area back to something like its presettlement condition — a condition Floridians regarded not long ago as fit for nothing at all.

We walk out on a parcel of restored saw-grass prairie to a study site where Hartman and her students recently planted 200 *Melaleuca* seeds, along with some seedlings and between 200 and 500 Australian pine seeds. Hartman had hoped to test how well the exotics would grow under different conditions. In the Everglades, soil types and water levels can change markedly within a few hundred yards. But the study plots are empty except for a few shoots of native grasses. "This was supposed to be our driest site," Hartman sighs, "but we had a decimeter of standing water here. You can imagine what our 'wet' sites looked like.

"We put seeds in fertilized plots, in unfertilized plots; we left them in the sun and gave them shade. We tried to get the whole range. Almost all of them died. We had so little success that we can't even draw any conclusions."

Hartman is particularly bothered by two implications of her findings. First, the U.S. Department of Agriculture tests species from abroad in greenhouses and control plots to see if they can be safely imported. "But if we can't propagate the species that we already know are problems, why should we think the tests they're running are valid?" she asks. "What about plants that are being created by bioengineering? How can anyone really tell what a species will do once it's released into the wild?"

Second, the historic flow of water through the Everglades, most of which has been siphoned off for development and agriculture, is in the process of being restored. If *Melaleuca* is indeed an aquatic plant, as Hartman suspects, the renewed water flow will create a vast new habitat for it. Wide buffer zones that are known to be free of *Melaleuca* have been established on the east edge of the park. But if the species does manage to invade the freshly submerged glades, it could overrun thousands of acres. "No one's come up with a good combination of herbicides that can be used safely to kill *Melaleuca* in standing water," she says.

Hartman also takes me to a boggy spot with a pretty array of grasses and forbs. Within a square yard are perhaps a dozen different species. "These are all native plants," she says. "This is the kind of recovery you get when you take the agricultural rock out.

"Every plant you see in here that's taller than a foot is dispersed by wind. Wind is absolutely key out here. Which is why there's so much concern over exotic plants getting blown in."

IT IS difficult for people outside south Florida to comprehend not only the property damage but the strangeness and community dysfunction caused by Andrew. Roads were impassable. Even after they were cleared, so many street signs and landmarks had been destroyed that drivers could not find their way around town. Some neighborhoods did not have electricity for three months. Bulldozers mistakenly knocked down houses and patches of woods that should have been left standing.

Monkeys and other wild animals escaped from the Miami Metro Zoo and dozens of private collections and wandered through the wreckage. A six-year-old girl discovered a lost cougar beneath a bush in her backyard. In the midst of such turmoil, the conservation biologists in charge of safeguarding south Florida's natural areas were forced to wait up to a year for the money, supplies, and equipment they needed to control the spread of exotic species.

One hot morning I explore the dark interior of the fifty-five-acre hardwood grove at Castellow Hammock Park, one of Dade County's showcase natural areas. During Andrew most of the hammock's large trees lost limbs or were knocked over, leaving huge rents in the vegetative dome. The understory was beaten down by the collapse of branches and trees. "It was depressing," Sandra Vardaman Wells says. "The whole place was a mess; you couldn't even walk a few feet into it. And it had the weirdest feel. Small things weren't right. All the tree snails were at eye level instead of up high in the branches."

Since September 1993 Vardaman Wells and a field crew have cleared debris from the ground and cut invading vines from the canopy, working on fifty-by-fifty-meter plots, one at a time. At first they attacked thickets of Brazilian pepper, jasmine, and wood rose that grew more dense each week. They propped up native hardwoods that had been knocked down but were still alive. Later they surveyed areas for exotic invaders, pulling up seedlings and spraying mature plants with herbicide. A few months later they went back and surveyed the areas again.

Now, four and a half years after the hurricane, we crouch in shade beneath a miniature canopy. Young hardwood trees form a tight cover, shielding us from the intense Florida sun. It is wonderfully cool and damp in here. A soft, buffered wind sends drops from an early morning rain onto our heads and backs. Through a gap in the branches turkey vultures appear and disappear, as silent as spirits. The canopy closes only five and a half feet above the ground — a few inches over Vardaman Wells's head — but with time it will extend upward. Beneath the tallest trees the understory is open, as it would be in a mature hammock, and seedlings poke through thick leaf litter. "It'll probably be thirty years before this hammock is back to the way it was," Vardaman Wells says, "but at least this is a start."

Vardaman Wells works with Joe Maguire at the Natural Areas Management Division of the Dade County Park and Recreation Department. In the wake of the hurricane the agency received $5.4 million in grants from the state and private foundations to restore the forested preserves that had been so badly damaged. "People knew that if something wasn't done to save the natural areas, they'd be gone in no time," Vardaman Wells says. In the past five years the county has spent half a million dollars to restore this one park.

We leave the grove and walk down a nature trail deeper into the woods. Reddish roots snake across the path. In places pinnacle rock juts through the dirt, looking like calcified Swiss cheese. The islands of hardwood trees, known as tropical hammocks, tend to grow on rock that is slightly harder and higher in elevation than that supporting pine forests. On this old limestone formation, a few inches of change in elevation can make a vast difference in vegetative cover.

We are deep in the subtropics, where life is lush and hardy and prolific. Wherever I look I see mahogany-colored branches bearing fleshy green leaves. Above us the burnished red trunk of a gumbo-limbo leans into the forest. In here the stark, sun-washed streets of Greater Miami seem as remote as another continent.

So do the temperate forest species I know from the places I have lived. As she walks down the trail, Vardaman Wells keeps breaking off leaves and crumbling them to release their scent. One smells like licorice, another like a sewer. For her, smell is an important clue to plant identification. The leaves of the various species all look distressingly similar to me, but Vardaman Wells points out subtle differences: drip tips, rounded lobes, wavy edges. Their names are lyrical: lancewood, paradise tree, pigeon plum, satin leaf. I admire her knowledge of this tiny botanical wilderness.

We move through the forest, talking. Biologists in the Everglades have gently accused Vardaman Wells of overstating the plight of the county's nature preserves, but she retorts, "They work with a natural system that's basically intact. All we have are little remnants. The devastation after Andrew was every bit as bad as we were afraid it might be."

She takes me to a sunny grove filled with lanky potato trees, which have slate-green, velvety leaves and orange fruit the size of large marbles. Years ago, after a fire set by arsonists, the ground where we stand was invaded by Boston ferns that grew to a height of six feet. The image of a marauding houseplant strikes me as surreal. Only in south Florida, I think. "Potato tree is one of the first successional species," she says. "Hardwoods will grow up pretty quickly and shade them out."

Farther into the hammock she scrambles down a rocky incline into a ten-foot-deep solution hole. I follow, and find myself standing in a rock-lined cylinder beneath the thick trunk of a fallen tree. The hole is about fifteen feet wide, with roots from a strangler fig dropping into it like ropes. I feel as if we are in a jungle far removed from the North American continent. The odd, tannin-stained rock that lines the depression has irregularly curled edges; it reminds me of iron devoured by rust. But it is too hard to chip or bend and is spottily covered with moss and ferns. Scant light filters in. The fallen tree and the vines keep the hole in perpetual shade. "I love it down here," Vardaman Wells says. "At one time this would have been filled with water, but the water table has dropped so much that most of the solution holes are dry.

"Hammocks with solution holes used to be everywhere," she adds, "and people took them for granted. Now, with all the development, there are hardly any left. What is it worth to save them? What are people willing to pay to have the kinds of hammocks and forests that have been here for thousands of years?"

It is a poignant question. Andrew was the first major hurricane to assault south Florida in thirty-five years, but major storms usually come ashore every fifteen to twenty years. Before Castellow Hammock and the Deering estate and all the other native oases completely recover, they may well suffer through another tempest of equal intensity. Biologists know the natural communities of south Florida can no longer survive without maintenance programs tailored to the size of the area under assault: the smaller the preserve, the greater the heroics that will be required to save it. But even in the Everglades, biologists will have to remain vigilant against exotics. The wind is simply too adept at dispersing aliens.

I leave south Florida a few days after my visit with Vardaman Wells. As I wait at the airport baggage check, in a crush of people babbling in different languages, I revel in the rough textures created by a meeting of cultures. How sad that all trends are toward homogenization, toward bland Westernization. I think of Vardaman Wells's words. What will it take to salvage diversity? What are people willing to give up; what are they willing to pay?

*

BACK ONCE MORE on the slender islands I call home, I find myself haunted by images of Andrew. I wonder what a hurricane with 145-mile-an-hour winds might do here, not to the dwellings and shops but to the natural landscape. A storm of such magnitude has not hit the Outer Banks for many decades. We are more fortunate than south Florida; we do not have as many invading plants. Still, there is no way to know how maritime forests and marshes stressed by encroaching development might be further damaged by a major storm.

On the Outer Banks, and all sea islands, vegetation is devoured not only by gales but by wind-borne salt. The harshest conditions lie at the ocean's edge. The only plants that grow on the dunes fronting the beach are grasses and low forbs with stems and roots that can flex and creep in blowing sand. The grasses' ability to bend makes them resistant to tearing, an important consideration in a natural system where wounds on stems and leaves can cause lethal uptake of salt.

Shrubs and cedars become established only where tall dunes provide shelter from northeasters, which aside from hurricanes are the stormiest, most salt-laden winds. In the few places where cedars protrude above the dune line, their skeletal branches all point southwest, like a quiver of spears. Look at any plant or bush on a barrier island, and you will likely find at least one feature — the height of its crown, the splay of its limbs, the cant of its trunk — that has been sculpted by wind. Several hundred feet inland, along the spines of the islands, grow compact live oaks with a stubble of dead branches on their windward sides. They are more like bushes than trees, and viewed from the side they remind me of ocean breakers. They rise gradually, their northeast edges sloping upward to a rounded peak, then roll abruptly down on the western side. Salt spray prunes them into this shape by burning off new shoots. The same phenomenon occurs with other island species, including wax myrtle and yaupon holly: over time, blowing salt molds them into a dense, streamlined form that is extremely resistant to wind. Deep inside the bushes, the younger, more protected leaves are larger and greener. Nonetheless, their tough cuticle loses moisture slowly, giving the trees an extra measure of protection from sun, heat, and blowing sand. Even on Roanoke Island, several miles west of the barrier beaches, the growth of pines reveals the harshness of island weather. In our yard the loblollies do not stand straight. Each one tilts slightly to the southwest, as if it had been leaned on all its life by wind.

On blustery days I sometimes seek solace in Nags Head Woods, a maritime forest that owes its existence (or so biologists believe) to a ridge of high dunes that buffers it from salty winds. On the northeast side of the forest, a sixty-five-foot dune called Run Hill deflects wind upward like a giant shield. Southwest of the hill grow beeches, maples, dogwoods, ferns, and tiny orchids, a whole lush assemblage of plants that do not belong on a barrier island. Marbled salamanders hunt in damp tunnels beneath fallen trees. Parula warblers fashion messy nests of Spanish moss. Standing on a ridge just below Run Hill, looking downslope into a soft array of leaves and ferns and flowers, I often imagine that I am not on the Outer Banks but in a hollow deep in the Appalachian Mountains.

Nags Head Woods is an example of how the movement of air across plains, up hillsides, and down into valleys helps create small pockets where weather conditions differ markedly from surrounding terrain. Such microclimates can be recognized by their unusual plant cover. A stand of misshapen conifers on a north-facing mountain ridge is a sure sign of high winds, while a grove of dogwoods on a southern slope reveals a protected, sun-drenched niche. In some situations growing conditions can vary markedly within a matter of yards. In medieval times, peasants in windy regions of Europe planted their fruit trees next to stone walls. The walls both shielded the trees and radiated heat that helped set the fruit.

In regions where trees are grown for timber, forests must be thinned carefully or the trees left standing will be more vulnerable to windthrow. The flow of air across a canopy — not just a forest canopy but one of crops or even grasses — tends to produce eddies that pass down through stems and stalks and trunks, causing the plants to sway and perhaps brush against their neighbors. On a windy day the passage of eddies can be seen in the rhythmic waves that skim across, say, a field of wheat. In a canopy with large gaps, the wind is able to swirl lower and inflict more stress on trees and plants than in areas of densely packed growth. Large gusts can intensify eddies; so can hilly terrain. Layers of wind are squeezed together as they travel up a hillside. But on the far side of the hill the squeezing is suddenly relaxed, causing a backwash flow that contains great turbulence and great potential for damage.

Exposure to strong wind may affect virtually every aspect of a plant's life. When a tree begins to grow, it stretches upward in search of sunlight; but if it becomes too lanky its trunk will not be substantial enough to withstand the stress of wind. A balance must be reached, and recent scientific papers contain references to the "design limits" of trees, as if the dimensions of each organism were carefully planned. In fact the "design" of a tree takes place as it grows. The stresses caused by swaying in the wind seem to promote more compact growth. Listing from wind may affect the slant of the wood grain in the trunk, for instance, and it may stimulate the release of biochemical compounds that govern growth. In addition, trees in windy areas depend heavily on their roots for anchorage. Wind tunnel studies showed that Sitka spruces and European larches exposed to extreme winds had 60 percent more root growth on their upwind sides, as if the trees were consciously seeking to grip the ground more firmly.

Forests may be as profoundly affected by an absence of breeze as by constant gales. In the rain forests of the tropics, the tallest trees form a tight, wind-brushed canopy, and the humid air on the ground stirs little. Instead of employing wind to disperse their pollen and seeds, rain-forest plants depend primarily on transport by animals and insects, and so have evolved showy flowers and luscious-colored fruits to attract feeders. This method also solves a problem posed by the great diversity of species within the tropics. In temperate climates, substantial numbers of the same plants tend to grow within windblown range of each other. As a result, when the trees release pollen on the wind, there is an excellent chance that at least some will fall on their own kind. In tropical rain forests, however, "if the traveller notices a particular species and wishes to find more like it, he may often turn his eyes in vain in every direction," wrote the great nineteenth-century naturalist Alfred Russel Wallace. "Trees of varied forms, dimensions, and colours are all around him, but he rarely sees any of them repeated." It makes no sense for rain-forest plants to manufacture quantities of pollen — an expensive undertaking in terms of the energy required — in order to have it broadly scattered by the wind.

In wind, writes the British author Xan Fielding, "trees are given tongues, and every species has its own voice." Even on days when I am locked tight inside my house with the windows shut, music playing, and a bubbling of conversation, I am conscious of the presence of wind because of the voices it gives to the trees. Air dragging through leaves and needles, past branches and limbs, has a sound unlike any other, a sound that changes slightly with locale and that — for me, at least — is capable of evoking powerful memories of place.

I thought of Fielding's image again and again the autumn I lived on the edge of Virginia's Blue Ridge in a landscape of tall hardwood forests and comparatively mild wind. On the rare days that the air blustered, it did so in brief bursts. During afternoons of moderate breeze, dried poplar leaves clattered to the ground like brittle tiles. Trunks creaked and branches whooshed. The wind had a noise and heft to which I was unaccustomed, all because of its effect on the trees.

One evening near Halloween, when the leaves had passed the peak of their color and about half had fallen, an unexpected gale came through. I sat alone in my starkly lit kitchen, my heart pounding. The rumbling of the wind was as foreign to me as a strange, hostile language; all night I worried that it might take the roof off. For those long hours the season of ghosts and goblins lost its jocularity. It seemed as if spirits of ill intent had been loosed on the town.

The next day, in calm, clear weather, I learned that several buildings had collapsed in Baltimore, downwind from us. The air speed outside my door had reached a paltry thirty-six miles an hour. The storm destroyed nothing near me but fanned more flame into whatever leaves still clung spottily to the trees. For the next week I reveled in a new flush of golds and crimsons, a fresh edging of orange.

To this day, wherever I go I notice how plants have adapted to wind. On the eastern slope of the Colorado Rockies, on a ranch owned by a friend, I pause to examine the contorted branches of a short pinyon pine. They are nearly as pliable as the limbs of a Gumby doll. Their rubbery nature enables them to withstand the brutal downslope winds known locally as chinooks.

Outside Livingston, Montana, where southerly winds hurtling up from Yellowstone reach some of the fastest speeds on the continent, I find ranch fields filled with little but short, wiry grasses — due as much, perhaps, to the presence of cattle as to wind. Cottonwoods and willows grow only in the moist, sheltered ravines of creeks. On the cool coast of northern California, ice plants hug rocky inclines above the ocean, tinted red by sun, their thick skins and succulent flesh well able to hold the moisture that would otherwise be sapped by salty spray. Wind's signature stretches from coast to coast and on around the globe.

THE ZOOLOGIST and physicist Steven Vogel considers Durham, North Carolina, where he lives, to be one of the most windless places on earth. "It's the exact opposite of the Outer Banks," he tells me when I visit him at his lab at Duke University, "but we do get an occasional big storm." And storms sometimes topple trees.

In the article "When Leaves Save the Tree," Vogel explains that while virtually all of a tree's weight is in its trunk and branches, leaf cover accounts for most of its surface area. This creates an interesting paradox when it comes to wind. In many deciduous trees the leaves stretch out horizontally to capture the most possible energy from the sun, to conserve water, and to aid the tree's cooling. The weight of the tree's wood and the grasping power of its roots all act to anchor it to the ground. But the drag created by its leaves may prove so strong during a storm that the tree topples, taking with it a sizable root ball. Uprooted trees, Vogel notes, are almost always in full leaf. Drag is greatest on trees with trunks that flex and sway. A stiff trunk is a more effective counterweight.

In the 1960s Vogel conducted a series of experiments showing that lobed leaves can cool themselves most efficiently in light winds, because air passing through the lobes forms eddies and brushes more of their surface. "It was an interesting study," he says. "It turns out that lobed leaves don't droop the way unlobed leaves do." In a related study, Vogel placed temperature probes on the undersides of leaves to measure how quickly they heat and cool. "They'd fluctuate five degrees centigrade in a matter of seconds," he says. "In Fahrenheit that translates to a minute-by-minute swing of thirty-five degrees."

Much later, Vogel began to wonder whether some trees had developed certain techniques for lessening their drag in high wind.

The twenty-five-foot-long wind tunnel at Duke University is not the high-tech mechanism I expected, but a simple metal chute with an interior test chamber five feet square. On one end is a fan driven by a forty-horsepower motor; on the other end is a grate and a set of squarish bellows to straighten the flow of air from the fan. "Most wind tunnels don't blow, they suck," Vogel says. "The air is extremely turbulent when it comes out of the fan. But that's the kind of flow I wanted when I started testing leaves."

For his experiment on leaves in high wind, Vogel cut small branches from a variety of trees and rushed them into his lab so he could test them while they were fresh. Arranging them just beyond the fan, where they would catch its full power, he observed and photographed them at wind speeds of eleven, twenty-two, thirty-three, and forty-four miles an hour. To his surprise, he found that as the wind increased many leaves became compressed into distinctly streamlined shapes. Tulip poplar leaves first folded down the broad lobes nearest to their stems, then curled up lengthwise. At twenty-two miles an hour they resembled paper airplanes in shape. As the speed of the wind rose, the tulip poplar leaves pulled more tightly inward, until they formed slender, stable cylinders with very little drag.

Holly leaves flattened against each other, making a thin stack. The compound leaves of black locust, which are arranged on long stems, curled against each other in long, slender tubes. The leaves of white oaks, on the other hand, remained almost completely horizontal. Vogel wondered if this species might rely on the greater weight of its wood to stay upright so its leaves could continue to absorb sunlight. But when he tested a larger white oak limb with many leaves, he found that the leaves clumped together in a mass that was much more stable and less vulnerable to drag than were fluttering individuals. "There may be other ways leaves reduce drag that can't be easily observed," he says. "Who knows what kinds of adaptations trees have developed?"

I drive home from my meeting with Vogel deeply distracted, stopping to gaze up at trees, determined to see leaves folding themselves into tubes. Through the towns of Wake Forest and Williamston, Plymouth and Columbia I watch a twenty-mile-an-hour west wind ripping through the crowns of maples, oaks, hickories, ornamental firs. The leaves of a white oak stay stretched toward the sun except in the hardest gusts, when they press together in a messy clump. At first I have difficulty telling what the firs are doing because they sway so much. After several minutes I see that they fold their branches backward, away from the wind, and hold their needled fingers straight out, like banners.

The strategy of the hickories is more obvious. Each compound leaf becomes a streamer, with six to ten leaflets folded around the stem, forming the compact tubes that Vogel described. An origami of wind. The maples and tulip poplars press their leaves together in clumps that are vaguely cone-shaped, but I can't tell what individual leaves are doing. All I see is a massive bending and nodding away from the wind, whole branches moving with the agility of human hands and arms, folding in toward the body of the tree.

It is a week full of wind. I am still watching trees two days later, when I drop our son off at preschool. In the parking lot another mother buttonholes me to talk about an upcoming fundraiser, a spaghetti dinner. She is saying something about desserts and how they need to be packaged for take-out, when a gust comes out of the northeast. High above us I see a silver maple press its leaves against each other, their whitish undersides shining. The wind pushes them roughly aside and travels on.

The other mother stops talking and follows my gaze to the crown of the tree. "Is something going on up there?" she asks.

NEAR THE NORTHEAST edge of the United States is a terrestrial anomaly that squeezes oncoming air like a set of mammoth bellows. As wind drives east over Vermont, it dips to scrape the valley of the Connecticut River and rises abruptly to crest the towering Presidential Range. It pours up the mountainsides, compressed into less and less space. "It's like putting your finger halfway over a water hose. The extra pressure causes the water — or the wind, in this case — to just go shooting out," says meteorologist Ken Rancourt. The bottom edge of the flow bumps against the mountains. The top edge hits the dense atmospheric layer known as the tropopause. As the wind rises to 6,288 feet, it passes over the summit of Mount Washington, long the site of the highest recorded wind speed on earth.

On a lovely summer day our family boards a car on the Mount Washington Cog Railway for a slow, jerky ride to the top. Jeff and I would much rather hike to the summit, but at age four Reid's short legs would never make the climb, and we do not want to carry him. So the cog railway it is. Reid perches on Jeff's lap, eyes wide as the steam engine lets out a belch and lurches forward.

With its chill temperatures and desiccating winds, Mount Washington is one of the harshest environs for plants outside the Arctic. At the summit the wind exceeds hurricane force (seventy-four miles an hour) four days out of ten. In winter, rime ice, formed from supercooled fog, plasters buildings, vehicles, and radio towers with a thick, bumpy coating. Only the hardiest moisture-hoarding plants survive here. As for animals, the most common by far is *Homo sapiens,* although a few mammals — among them shrews, voles, red foxes, and long-tailed weasels — migrate above the tree line in warmer months. Clouds live at the summit, and rocks, not wild creatures. The Abenaki Indians who ventured up the mountain called it Agiocochook, the place of the storm spirit.

As we climb we will pass through increasingly frigid country. In terms of climate, every 1,000 feet of elevation we gain will have the effect of carrying us 250 miles farther north. Leaving from Bretton Woods, New Hampshire, it will be as if we have been transported to northern Quebec.

The locomotive begins its ratcheting ascent in a pretty forest, with ferns and wildflowers edging the track. Soon, though, the deciduous species thin to mountain ash and white birch, and conifers appear. The lush and frail have been weeded out by wind and cold; there is no room on the higher slopes for plants that lose moisture quickly through their leaves. Botanists divide high mountain country into three ecological components: the northern hardwood zone at the base, the spruce-fir zone, and the arctic alpine zone. The plants within the lower two belts arrange themselves by height, growing shorter up the mountainside as the temperature drops and the speed of wind increases.

The train chugs laboriously; the bitter smell of burning coal swirls through the car. On the mountainside birch and ash diminish to six feet, and their branches grow twisted and flagged. I turn briefly from the window to answer a question from Reid. When I look back, the hardwoods have been replaced by black spruce and balsam firs with slim trunks growing close together. By 4,600 feet fir branches form an interlocking mat that obscures the ground. "Look," Jeff says. "It's like a solid fabric." A canopy from wind. I wish I could jump off the train and explore what lies in the protected niches below.

Higher, and the trees thin, interspersed with patches of open meadow. The conifers slouch crazily upslope, bent low like soldiers overrunning a hill. Behind isolated boulders they dare to grow a bit taller — perhaps all of four feet — but their top needles are burned brown. This dwarfed, bonsailike growth is called krummholz, and it colonizes hillsides where snowfall offers an insulating blanket or where uneven terrain provides scattered places of shelter. Krummholz is a German word meaning crooked wood. It becomes established when a seed is blown upslope and finds a niche, say in the lee of a boulder. As it matures, the tree grows bushy and shrub-like. On its lee side new branches sprout and creep uphill until conditions become too harsh for further growth. Lacking the energy to produce seeds, the dwarfed trees spread by sprouting roots wherever their branches curl down to a bit of soil. On Mount Washington, krummholz has been dated to one hundred years. In the less severe conditions of the Rocky Mountains, it may live more than three hundred years and be found as much as 6,000 feet higher.

Tree limit is imposed by a number of factors, including climate, snow cover, soil, and competition from other plants, but perhaps the most important is exposure to wind. In winter, when soils are frozen and water is unavailable, high winds parch trees mercilessly, and blowing snow and ice may damage their needles and bark. Where the flow patterns allow tendrils of cold air to reach far down mountainsides, the tree limit lies at lower elevations. On leeward slopes, though, the trees creep upward, finding small microclimates where they can survive, if not thrive.

As we reach 5,000 feet, firs sprawl uphill in gray-green lumps a few inches high, the thin edge of an ocean of vegetation. On the mountainsides to the west, prostrate conifers lap upward like foam blown onto a beach. A gust blasts through the open car, chilling us as quickly as if someone had thrown cold water on us. Reid hunkers against Jeff's chest. The train lumbers at a steep angle, chugging through fields of wiry grasses and lichen-covered boulders to the balding summit, where clouds pass low overhead and several hundred visitors wander among rocks.

IN 1932 the privately run Mount Washington Observatory opened with an anemometer on a cast-iron shaft that could record the gusts in what was already suspected to be one of the windiest locales on earth. The anemometer measured wind speed through a series of clicks that recorded its revolutions. On April 12, 1934, a day of unusually strong wind, a technician named Sal Pagliuca ventured outside long enough to chip ice off the anemometer and other instruments. "Perhaps a sledge hammer could have done a better job," he wrote in a journal, "but I doubt if the strength of Polyphemus could move a sledge hammer in a 200 mph breeze." Retreating inside, he settled down to count the anemometer's clicks. He had never heard them coming so fast. Working with a colleague, Pagliuca figured the wind velocity at 231 miles an hour. "Will they believe it, was our first thought," he wrote. "I felt the full responsibility of that startling measurement."

Other people did believe it. The observatory held the world record for wind speed until December 1997, when a typhoon that ravaged Guam generated gusts of at least 236 miles an hour. But anyone who has been to Mount Washington's summit should not be surprised that the mountain long claimed the world record. Standing on the terrace of the visitor center, we watch clouds only a few yards over our heads (or so it seems) being pulled to tatters by wind. Far above, stately cumulus towers parade by much more slowly. Today it appears that the wind is moving more quickly close to the ground. But no detailed studies have been done of wind flow through the Presidential Range, so it is impossible to guess how currents may be colliding and diverging over our heads. Recently Ken Rancourt of the observatory told me that meteorologists are not even sure why the wind builds to greater speeds here than it does in the surrounding region.

Today's mild breeze of fifteen miles an hour has a biting chill, and we seek shelter inside the bustling visitor center. I hold tight to Reid's hand to keep from losing him in the crowd. We find our way to a corner where a video monitor is showing a continuous film clip called "Breakfast of Champions." A diner arrives at the fictitious Mount Washington Cafe and asks for a terrace table. The host tells him it's a bit windy. "That won't bother me," the diner replies heartily. The next scene shows him seated outside in a full winter survival suit, holding a card table down with both hands. The edges of the tablecloth snap like a flag. The diner's coffee arrives, but the cup goes flying out of his hand; so does a slice of toast. A waiter arrives with cereal and milk, but when he tries to pour them into the diner's bowl they stream off like talcum powder. The diner's chair tips over, spilling him to the ground, and the table blows away. Reid shrieks with laughter. The speed of the wind? A mere sixty-five miles an hour.

I leave Jeff and Reid briefly for a visit to the observatory, where I talk with a technician named Norm Michaels. The observatory occupies one wing of the visitor center, a blocky concrete building constructed in 1980 and designed to withstand the most severe storms. Weather instruments are mounted on a tower that rises above the terrace. As Michaels types his hourly instrument readings into a computer, I look west out a picture window to the cloud-dappled forms of Mount Jefferson, Mount Adams, and Mount Madison.

Michaels has worked at the observatory for seven years and has seen the most extreme weather Mount Washington has to offer. "You should have been here last month," he says. "We had winds to one fifty." A low-pressure system moved off the New England coast and then north over Nova Scotia, stretching cold air across the mountains. "It was extremely unusual for the middle of the summer. Look here." He shows me a wall where a Hays' Chart Recorder hangs, mechanically tracking the wind speed on a round disk. Today's wind registers as a series of mild peaks and dips that reach less than a third of the way across the disk.

"Now here's the chart from July 20," Michaels says, handing me another paper circle. It looks something like the spin paintings I remember from a childhood art class. Ink is sprayed heavily across it, reaching to within an inch of the margin. Slashing lines at its outer edge show extreme gusts. "Quite a difference," Michaels says. "But I've got one that's even better." From a file he pulls a chart made in December 1980 on a day when gusts were measured to 182. Ink covers it almost completely. "That was the windiest day we've ever had in this building.

"Let's go downstairs," he adds. "I've got a video you'll find interesting."

He leads me down a circular iron stairway to the quarters where the observatory staff live during weeklong shifts. In the living room Michaels roots through a box of videos and pulls out a cassette. "Here's some footage of that wind we had in July," he says. "The gusts were so strong that they pulled up some tiles on the terrace. Pretty amazing when you consider that the tiles each weigh seventy-five pounds. One of the other technicians went out to try to put them back in place."

The footage lasts only a few minutes. At first the camera shows Steve, a large man, edging his way along the terrace railing, his features obscured by a survival suit and goggles. Rain slides sideways across the screen. Steve moves tentatively toward the tiles and crouches long enough to scoot one back into place, straining beneath its weight. To reach a second tile he must venture out of the lee of the observatory and into the full force of the wind. The fabric of his suit heaves and whips. Wind roars in the camera microphone. Steve takes a step, moving quite slowly, bracing himself, but he is immediately knocked to his hands and knees. He rises ponderously and begins inching back toward the camera. A gust catches him, and he tumbles and rolls down the observation deck. The last shot shows him crawling back as rain sweeps around him.

OUTSIDE AGAIN Reid and Jeff and I decide to hike a short way down the east side of the mountain, which today (and on most days) is protected from the wind. Although it's the middle of the week, the trail is crowded with people climbing up. Mild weather is fleeting on Mount Washington; even in midsummer cold fronts can move in so quickly that hikers are always advised to carry wool sweaters and down vests. On warm days like this the trails are heavily used.

In her book *Land above the Trees,* Ann Zwinger writes of a visit to Mount Washington on a summer day when clouds obscure the summit and the world seems to be made of only two colors, gray and green. But, Zwinger notes, the colors look as if they have been assigned to the wrong forms. The rocks, encrusted with map lichens, look like "mottled green marble. The plants are so cloud-caught that they appear gray. Green rocks, gray plants: a Norse landscape." Further on she adds, "The mists and fog and vapors and rain devour shapes and forms, bleaching all colors gray. They consume the landscape like an amorphous, ravenous, ice-breathing dragon."

The hues that greet us on the eastern slope are quite different from what Zwinger saw that cloudy day, but their odd arrangement makes me slightly dizzy. From afar the rocks appear to be spattered with fluorescent paint. Moving closer, I see that they are thinly covered with black-and-green lichens. Against the deep blue sky of this altitude such a neon green seems weirdly out of place, but it speckles boulder after boulder upslope and down.

We pick our way off the trail for a short distance, warily watching Reid, who suddenly believes himself to be a great mountain climber, incapable of being hurt. His legs look extraordinarily long to me, and he pulls himself over rocks at a speed I never thought possible. Jeff rushes to keep a step or two behind him. I move more slowly, exploring the mountainside a few feet at a time. Rock edges scrape my bare shins and thighs.

In between the boulders I find a riot of textures. Grasses, lichens, mosses, and tiny herbs crowd into crevices where bits of soil are exposed, forming a tapestry of plant matter, a cloak of many colors. The grasses are four to five inches high and silky, with very thin stems. The mosses look spongy and welcoming. I touch a deep green cushion and am surprised to find it brittle. No soft edges allowed, not up here at the limit of life. A few feet away grows a brownish patch of Iceland lichen, also pillowlike in appearance but stiff to the touch. Its minuscule leaves flute and curl in every direction. I pull up a plug of it. Just below the intricately branching tips, the reaching, gray-white stems remind me of the cauliflower arteries within the human brain. It is a form I have seen many times in nature: a tight, protective canopy acts as a shield for the growth of larger stems below. In this case, though, the canopy reaches a mere two inches above the ground.

A number of years ago I took a hiking trip in the alpine fields of the Rocky Mountains. Now, as I find flowers hidden between rocks like jewels, I remember my initial sense of awe at what nature can create in miniature. In the meadows and fell-fields of the Rockies I discovered alpine forget-me-nots, each bloom less than an eighth of an inch in diameter, and dwarf clovers, and snow buttercups. Rock jasmine, moss gentians. I spent whole afternoons on my hands and knees, combing through natural rock gardens. The plants grew in rounded cushions and mats shaped to deflect the wind.

I have missed the height of wildflower season on Mount Washington, but a few blooms dot the rocky slope. I make my way carefully over to a yellow aster. It grows six inches tall in a patch of long, lush grass; there must be a source of moisture here. On the slope above it, grasses tumble down the rocks, wedged into crevices, a cascade of vegetation in a rocky desert.

We climb back toward the summit, stopping to look at a thick patch of diapensia, a flowering plant that prospers in areas so exposed to wind that no snow covers them even at the height of winter. Its needlelike leaves form an interlocking mat, their flat sides offered to the sun for maximum warming. In early summer diapensia blooms profusely, with small white, five-petaled flowers, but they have long dried.

Reid has reached his limit. No amount of sweet talk or teasing will convince him that we should stay on the mountain a minute longer. Even the belches of a coal engine with a bulbous stack cannot hold his interest. We join a lengthening line for the return train and watch the wind stretch clouds into gauzy scraps.

There is only one seat left by the time we reach the car. I give it to Jeff, who cradles our nodding son, and retreat to the back to stand on an open platform. As soon as we leave the summit, the wind-pruned grasses grow longer and shinier. Five hundred feet down, sprawled firs reappear in patchy hummocks. Lower, an intricate weave slips over the scaly mountainside, a coniferous cloth spangled with deep red berries and bright yellow flowers. Then the terrain grows marshy; sedges and bogberry stand in small pools. We descend into taller trees — a whole six feet high! — with a sense of relief. Light sifting through deep green needles is a balm to eyes worn out by sun.

Ferns line the track now like delicate green flames. As we come into full-sized forest, a tide of wildflowers washes below the train. Goldenrod, lobelia, Queen Anne's lace. I barely noticed them before. We are back in the world of the upright, the vertical, where lushness is the rule, and gloriously so. Hunched all day against wind, my back kinked from poking among plants no taller than my ankle, I stretch my arms overhead, throw back my shoulders, and stand.

SIX

The Waltzing of Wind and Sand

ON A HUMID EARLY SUMMER afternoon, two women in baseball caps and grubby, loose-fitting clothes pause from a labor of staggering dimensions to guzzle water from canteens. Above them the crowns of sixty-foot oaks and pines rock back and forth in a salty breeze. The wooded hillside where the women sit is strewn with shovels, root cutters, measuring tape, and a sizable mound of loose sand. At the center of all this are two rectangular pits, each about five feet deep. "Well," the shorter woman says brightly, surveying the pits, wiping her cheek on her sleeve, "we're getting there." With their dirt-smeared faces and sly smiles, the two look as if they are bent on digging to China.

It is not a place they seek to reach, of course, but a time. The more petite woman is Karen Havholm, a geologist who specializes in studying aeolian geology. Meaning, for the most part, sand and sandstone.

"New blood," she exclaims when I walk up. Her companion, an undergraduate geology student named Kristin Weaver, remains seated. They have been digging since seven-thirty, five hours now, with only a brief lunch break. I have offered to help dig in return for a geologic tour of Nags Head Woods. "What have you found so far?" I ask.

"No soil line, which is what we need to find for any of my theories to make sense," Havholm says. "But there are some interesting cross strata here." She jumps into one of the neck-deep pits, knocking a rivulet of sand from its side, and I follow.

The top two feet of exposed soil are held fast in a lacework of lopped-off roots, some as thick as broom handles, many more as thin as thread. Below is a wall of buffy compacted sand. Havholm points to the south-facing wall of the pit, to an odd arrangement of faint pastel stripes, pale yellows and subtle browns. Just below the roots the stripes run horizontally through the soil, with a slight upward tilt. These were laid down by the back side of a dune that was migrating slowly southward. Two feet lower, the upward-slanting layers abruptly end; here the stripes point steeply downward, disappearing into the ground. To me it looks as if blocks of earth have been stacked tightly against each other at crazy angles.

"These lower, dipping stripes show where the dune originally moved through," Havholm says. "Each stripe represents an event, like a grain flow down the face of the dune. Then the back of the dune came through and laid down the top laminations.

"I was hoping to find some sort of boundary laminations between two separate dunes," she adds wistfully. "But I'm beginning to wonder if we shouldn't be digging farther over into the side of this hill."

"Don't say that," Weaver admonishes. The tree-covered slope rises so steeply that the layers of humus and decayed roots Havholm seeks could be twenty feet below the surface.

Untold centuries ago, perhaps as long as five thousand years, the land beneath Nags Head Woods consisted of great roving sheets of sand with irregular, migrating dunes. At some point the march of sediment slowed and vegetation took hold. Ever since, the rises and dips of the land have been cemented in place by a pervasive stubble of roots. To cut a straight path through the woods, one must climb fifteen- and twenty-foot crests, like a ship in a storm-tossed sea.

Yet to those who can read geologic rhythms, the waves exhibit a clear pattern. A few days earlier Havholm had shown me a tattered aerial photograph of the woods and had traced for me the outline of several large, arch-shaped dunes known as parabolics. When they were active, Havholm said, the parabolic dunes must have migrated from north to south and collided with a series of smaller sand ridges. "The question is, which are older, the parabolics or the smaller, east–west-running ridges?" Havholm asked. "And how did the movements of one set affect the other?"

With the help of Weaver, her companion for three weeks of fieldwork, she has already dug three other pits along a power line that crosses some of the highest dunes in the forest. Now the two are probing the sand where a parabolic — the large, steep hill — intersected another ridge. All in pursuit of a few isolated clues to the terrain's past.

Over the next several years Havholm hopes to compile a detailed history of dune movement on this middle section of the Outer Banks. In probing the buried layers she hopes to find clues about why dunes form and begin to range across the landscape like ponderous beasts, devouring trees, meadows, houses, and whatever else lies in their paths. What triggers dunes to stand still long enough to be tethered by grass and trees? One factor is moisture. In humid coastal regions dunes tend to be cloaked in vegetation, but occasionally great pulses of sand bury the plants and start the dunes moving anew. "There must be some sort of sudden influx of sediment," she says. "The question is, where does it come from? A drop in sea level? A radical change in climate? There are probably a number of complex factors." Why else might dunes lose their protective covering and begin to roam? The answers to such questions have sweeping implications for the management of the shifting slips of land that hold so many of the world's vacation meccas.

It is a foolish person who builds a house on sand, as the gospels warn. Yet millions of us have ignored that advice and allowed ourselves to be drawn to the cool breezes of oceans and lakes as if to an intoxicating nectar. Once there we strive to cement our property claims with careful surveys and plats. Like all lines in the sand, these are destined to be erased by water and wind.

On the southern boundary of Nags Head Woods is a famous sand hill known as Jockey's Ridge, which once formed the highest dune on the East Coast — 110 feet — but has since deflated to a mere 87 feet, losing its title to the dunes of Cape Cod. Jockey's Ridge is the most southern of a series of giant dunes known as medanos that once dotted the coast south of Chesapeake Bay. Kill Devil Hill, where the Wright brothers first flew a powered plane, is another medano, but it has been stabilized by grass.

Jockey's Ridge and its nearest northern neighbor, Run Hill, are still bare sand, free to travel. And travel they do, in small, lumbering avalanches. The leading edges of these dunes move as much as five feet southwest in a single year. In the process, Run Hill devours grove after grove of trees in Nags Head Woods. Jockey's Ridge buries backyards, swing sets, and toolsheds and occasionally forces home owners to move their property out of harm's way.

In 1991 Havholm and a colleague were hired by the Nature Conservancy, which owns most of Nags Head Woods, to conduct a study of Run Hill. The terrain so piqued Havholm's curiosity that she put together the funding for a study of the region's dunes. "Most of the aeolian rock record is sandstone," she had told me. "Nags Head Woods is a pretty unusual study site — a coastal system with multiple dune phases — so I'm hoping to learn quite a bit about sedimentary processes in coastal areas."

Now she is cutting, shovelful by shovelful, into sand, fatigue etched across her face.

It is time for a midafternoon break. We leave the tools behind and hike the half-mile to the nearest road, swinging our arms, reveling in their lightness. "I hope all this is worth the trouble," I say.

"You can never tell," Havholm replies matter-of-factly. "That's why geology is such a labor-intensive business."

We follow a path just north of the camel-backed parabolic dune. Along the way Havholm points out several smaller ridges that taper off the parabolic like wide-flung arms. Earlier in the week she and Weaver took corings from the ridges in hopes of finding variations in sand grains or, even better, a layer of humus. "A buried line of soil shows where a moving dune migrated over a vegetated dune," Havholm says. "The strata above and below the line are separated in age by at least a couple hundred years; the dune had to have time to vegetate and then be covered over again. So a soil layer can be pretty important in helping us sort out the big picture."

She stops walking suddenly, takes a compass from the pocket of her filthy lavender pants, and plows off the path through branches and vines. "She's probably gone to check the orientation of that smaller ridge," Weaver explains.

Through catbrier and holly I can see Havholm studying the compass in her flattened palm. Patches of light flicker across the scene like sun brushing the bottom of a pool. I am distracted by the malevolent, laughing call of a pileated woodpecker, the cheerful rattle of a pine warbler. It's clear that I am going to have to learn the fundamentals of a new discipline in order to understand Havholm's work. In past seasons I have hiked these woods in search of birds, butterflies, wildflowers, and trees — and now geologic forms. How many ways are there of looking at a landscape?

JUST AS IT has shaped the history of humankind, the seasonal paths of animals, and the spread of vegetation, wind chisels the crust of the earth. It whistles around mountains and through passes, eroding rock as it gains speed. Bit by bit it skims the tops off plowed fields. It scatters ash from volcanic explosions and so creates some of the richest soil on earth.

In the Mississippi Delta, the heartland of Iowa, and the plains of northern China and Mongolia, wind drifts tiny, puzzlelike crystals of a mineral mix known as loess into thick beds. As the crystals land, they lock together to form a rock as powdery as talcum but as hard as pumice. In the dry valleys of Antarctica's Admiralty Mountains, wind blasts pebbles with ice and sleet until they are polished as smooth as glass. In the American Southwest, it scrapes away the edges and corners of rock, forming arches and columns and palisades of deep red stone. And wind pushes massive dunes across the deserts of the world, just as it did millions of years ago, when a vast sand sea inundated what would later become Colorado, Utah, New Mexico, Arizona, Nevada, and Wyoming.

Until the late eighteenth century, geologists believed that rock structures were formed exclusively by sudden, cataclysmic events, such as the eruption of volcanoes and the thrusting up of mountains. By the turn of the twentieth century it was known that the formation of rock sometimes required millions of years. But even then scientists generally assumed that the earth's terrain was shaped mostly by water — rain, rivers, and expanding and contracting seas. The role of wind was not given its due until the 1930s, when the plowed soils of the American Midwest began blowing away by the ton.

In 1925 Ralph Algiers Bagnold, a young officer in the British Royal Engineers, was stationed in Abbasiya, on the outskirts of Cairo. Bagnold was a student of fluid mechanics and a careful observer. He took to exploring the Libyan desert with some other men, spending his occasional ten-day leaves driving through the trackless sands. Traveling the desert became Bagnold's passion. As often as possible he abandoned the city in favor of "the sense of freedom to go just where one liked, driving on a compass course; with the awe of a new and utterly lifeless world; . . . with the clean coolness of sand dunes in the evening, and the dry sparkling desert air." He returned to England a few years later determined to seek explanations for all he had seen. For the rest of his life he immersed himself in studying the elegant waltzings of wind and sand.

In 1941 Bagnold published *The Physics of Blown Sand and Desert Dunes,* which remains the primer for students of aeolian geology. He begins the book with an examination of the movements of individual sand grains as they are tossed and tumbled by wind. A grain first starts to move by rolling, becoming airborne only when it runs into a barrier of some sort — a pebble, a bush, a crust of hardened sediment. It hops into the air, lands, and is thrown upward again, bouncing across the landscape with bigger strides as the wind accelerates. This hopscotch is called saltation. Wherever the grain lands, it forms a tiny crater, and its impact sets other grains in motion. Bagnold conducted painstaking tests in simple wind tunnels to study the paths of individual grains and their behavior under different wind regimes. He found that sand grains bounce much higher when blown over a surface of cobblestones than when they pass over sand. He demonstrated that sand ripples always form at right angles to prevailing winds. And he showed that sand exerts force on wind, as well as vice versa. Once grains begin to saltate, the surface winds that set them in motion are slowed by friction.

Bagnold's fascination was not with coastal dunes, which are often "chaotic and formless," but with the great deserts, where wind and sand are unaffected by humans. With no barriers to block its passage, no vegetation to tie it down, sand sifts into globed and tapered piles that change daily, hourly. An observer, he writes, "never fails to be amazed at a simplicity of form, an exactitude of repetition and a geometric order unknown in nature on a scale larger than that of a crystalline structure." He celebrates dunes "lined up in parallel ranges, peak following peak in regular succession like the teeth of a monstrous saw for scores, even hundreds of miles." He wonders at the phenomenon known as singing, or booming, sand, "which in some remote places startles the silence of the desert." Its cause, he finds, is the avalanching of grains, which slow and moan as they reach the bottom of a steep slope. And he marvels that two large dunes sometimes "breed" by colliding, then rolling onward, leaving behind a new small dune. Such "grotesque imitation of life," he writes, "is vaguely disturbing to an imaginative mind."

The major dune shapes Bagnold describes — barchans, or crescentic dunes, which have steep slip faces and gradually sloped back sides; seifs, or sword-shaped dunes, which stretch forward in long swirls, like curlicue waves drawn across a page; and sand levees, or whalebacks — seem like species of animals as much as variations in landforms. Scientists have since added other forms, such as the parabolics, which migrate by pushing their rounded noses forward and tucking their arms behind them. Patterns of wind shape them all, with sure and beautiful strokes.

Bagnold's work stood virtually alone in the field of aeolian geology until the 1970s, when the formation of the Mideast oil cartel spawned a flurry of research into modern sand seas and the oil fields below them. During the same period, observations taken from the *Viking* and *Mariner* spacecrafts showed that windstorms have been instrumental in shaping the terrain of Mars. Although the field of aeolian geology has greatly expanded since the 1960s, it is still a specialty science. Havholm estimates that no more than two dozen scientists in the world actively study windblown sand. Some are probing the processes of sediment transport on Mars and Titan. A few are exploring the qualities of loess in Mongolia and north China, where whole cities have been carved into its thick, buffy beds.

Most aeolian geologists, however, study what they call the rock record of sand seas and sandstone. "Geologists work by analogy," Havholm says. "They examine modern processes and then look at places like Zion National Park and run the scenario backward. There's a phrase used to describe it: the present is the key to the past. The overriding assumption is that many of the same basic processes at work today were also at work when ancient sandstones were formed."

Some aeolian geologists study sand movement on the microscale, using time-lapse photography and computer models to analyze how ripples form under certain conditions, or how a combination of waves and wind carves out cusps along a beach. Others look at the behavior of single dunes and whole seas of dunes. Much of this latter work is done by computer analysis.

Eventually, however, all theories must be tested in the field, using the same laborious trenching techniques employed by Havholm. "There are a few tricks to making the work go quicker," she says. "You can take samples with a vibracore, which is basically a big shaft attached to a lawn-mower motor that drives it into the ground. But that works best in wet sediments. I haven't had much luck with vibracores in dry sand, and I don't know anyone who has."

Once a trench reveals the pale stripes known as laminations, the geologist must decipher them with great care. As sand rolls and saltates and slides, the dense minerals within it sort themselves into layers that can be easily seen. Each stripe may represent any of a number of things — a sustained wind ripple, a spot where sand adhered to a damp surface, or a grain flow down the advancing face of a dune. By examining grain patterns and doing various calculations to compare the angle of a stripe and the orientation of the trench walls, investigators can determine the direction in which a dune moved.

Such clues may elegantly reveal where a large, quickly migrating dune overtook a slower, smaller ridge. Or they may simply confound the researchers. "More than once I've been in the field with what I think is an ironclad theory," Havholm says, grinning ruefully, "and the cross strata have blown it completely out of the water."

Dig and dig, take notes and pictures, pause to reevaluate theories, to study old topographic maps and aerial photos. These are the tasks that fill Havholm's days, from 7 A.M. to 10 P.M., during her short weeks in the field. Often when I ask a question she replies, "I'm not sure. I haven't thought that one through yet." After spending a day with her I fall exhausted into bed. I dream of sand spilling in hot, velvet tongues down the side of a dune, spreading out in fans, accumulating grain by grain until it smothers the forest that stands in its path.

IN 1991, when the aeolian geologist Gary Kocurek, one of the field's preeminent scholars, and Karen Havholm examined Run Hill, part of the dune was owned by the county board of education, which planned to build a high school on its crest. Nature Conservancy officials feared that the construction would cause severe erosion on the remaining section of the dune. If the dune lost much elevation, it would no longer shield Nags Head Woods from maritime winds, and the forest would likely die.

The planned site of the school has since been moved, and proceedings have been started to bring Run Hill under the ownership of the North Carolina parks system. Metal rods have been driven into the dune in a grid pattern so Havholm can monitor changes in sand accretion. "There are two parts to what I'm studying: the history of the dunes under the woods and the status of Run Hill as a modern migrating dune," she says. "Run Hill has lost its sand supply. Most of the sand fields to the east that historically fed it have been stabilized by lawns and houses and driveways. So the question is, how long can it survive?"

It is late afternoon, and Havholm and I are standing on the crest of Run Hill, one of my favorite places on the Outer Banks. The sides of the dune are bone white, but scattered patches of dark and light sand mottle its top, like an Appaloosa hide. Overhead, clouds skate quickly by on a building southwest breeze. As the wind sets grains of sand bouncing, the line between air and earth blurs. Thanks to Havholm's tutelage I know that here at the top of the dune the layers of wind, squeezed together as they ascend, are picking up speed and sand. Once the winds pass over the crest and start down, they will spread out and drop much of their load.

Havholm has reached the end of her stay; she will not return until winter. She has still found no buried lines of soil nor any clues to help her definitively date the dunes. "The dates are key," she says. "Until I can figure out how to get some sort of a time frame, I won't have much of a study. We can't use carbon-14 dating in sand, and there's no pollen in the dunes that can be dated."

Despite this setback she seems freshly intrigued by the form of the dune. "Look," she says, pointing to a series of half-inch-high ripples in the sand. "This is a good illustration of aeolian processes in miniature, at least for transverse dunes." That is, dunes with crests that stretch perpendicular to prevailing winds.

"When you see an area like that," she says, nodding toward a patch of mottled sand with large grains, "you can be pretty sure it's deflating. The fine stuff from the surface layer has all been blown away. And if you see an area that's very soft and smooth, you can be pretty sure it's accreting."

With its steep slip face, Run Hill is considered a crescentic dune, though it is shaped like an amorphous blob. A few yards southwest of where we stand, the dune cascades steeply into the north end of Nags Head Woods, which it is consuming tree by tree. The dune both gives life to the woods and takes it away. Beyond the woods sweeps Roanoke Sound, with the familiar hump of Roanoke Island — my home — in its center.

The delicate, straight-file tracks of a fox meander up a crest and cross the deep, ugly tread of a Jeep that was driven illegally on the dune. "When we first did our study of Run Hill," Havholm says, "we didn't think all the trucks that come up here to joyride were doing much harm. Now I'm not so sure. If we could keep vehicles off, the dune might vegetate more quickly. That's the natural fate of it anyway; someday grass and trees will cover it."

"It's sad to think about that," I say. Run Hill and Jockey's Ridge are all that remains of the Outer Banks in their wildest state — pure, blowing sand.

"Maybe so," Havholm replies, "but at the rate things are going, there's not going to be much dune left if something isn't done to keep it intact."

I remember Havholm's words six months later as I begin a hike up the seaward side of Jockey's Ridge on a breezy winter day. The wind is due west and stronger than twenty-five miles an hour, but here on the eastern side I am sheltered. I look up to the top of the ridge, craning my neck. When a sand hill reaches a pitch greater than thirty-three degrees, its grains slide downward, or so Havholm has told me. Today Jockey's Ridge seems to defy gravity. With its vertical face and sloped shoulders it reminds me of Yosemite Valley's famous Half Dome, a sphere of granite sliced in two.

I climb quickly, my back hunched, my chin against my chest. As I near the summit a vertical fissure splits the soft sediment of the steep face. Cascading sand buries my feet and shins; I have to scramble hard to break free. On top now, I squint west across the platinum surface, using my hand to shield my eyes. I can't breathe for sand. I turn around and lean against the wind as if it were a wall. Over the ocean, clouds with ink-blue bottoms mass and shift. Their white forms, lit by late afternoon sun, look like a second, airborne layer of breakers. Around me sand spirals upward in smoky plumes.

Jockey's Ridge is a dear landmark to Outer Banks residents and visitors, and it is shrinking. I cannot help feeling that I am watching its slow death. The sand blown eastward today will be carried west again later in the winter and will likely end up in Roanoke Sound. Would it be best, I wonder, to freeze the dune in time by grassing it over? Should we deprive it of its fluid freedom, lest it silently blow away?

TO STAND in the midst of a sea of great dunes is to have the sensation of being utterly alone, and vulnerable, in a world of scorching elements and soul-shaking beauty. I have felt this many times on trips to the Oregon coast, where thousands of years ago an unusual alignment of rivers and headlands caused the formation of a sandy desert only a few miles wide but fifty miles long. It is what the dunes beneath Nags Head must have been hundreds, or perhaps thousands, of years ago.

On each of my trips I climb a forty- or fifty-foot dune, not always the same one; but I climb purposefully, as if partaking in a ritual. My legs ache in protest as I move upward, sand sucking at my feet, small avalanches burying my shins. I grab for the silver, fibrous roots of beach grass, struggling to stay upright. Halfway up I tire of my incremental progress and, though I know better, try to bolt — and find myself hobbled by sand, able to move only in comically slow motion. I give up and fall, laughing, chest first into the dune. Fine, sparkling grains stick to my fingers and palms.

When finally I reach the top I imagine I am on the back of a great whale, surrounded by a half-dozen other surfacing animals. Even in humid Oregon the sun on my shoulders feels as if it might suck every drop of moisture from my body. To the west I can see a thin line of the Pacific, a lucent blue against the sand. I crave water, rest, and shade. But more than anything else I want to run down the side of the dune, arms spread and whooping. And that is what I do. The wind erases my footprints within the hour.

To scale a slickrock wall beneath a piercingly blue sky, on the other hand, is to walk across the back of time.

One cool December afternoon I seat myself on a narrow ledge of orange rock — a bounding surface, Havholm would say — in an attempt to stay in the sun but out of a whipping wind. At my back is a mountain; in front of me is a classic desert panorama: columns and humps and mounded loaves of sandstone, burnt red with swirls and tongues of white. It is Navajo sandstone, the young rock that forms the monoliths of Zion Canyon, the foundation of Arches National Park. All of it was once a great sand sea, pushed and shaped by wind.

Two days ago Havholm and I met in Phoenix and drove north into the rock countryside surrounding Lake Powell on the Arizona-Utah border, where from 1986 through 1989 Havholm conducted the fieldwork for her Ph.D. dissertation. It is the week between Christmas and New Year's Day, and Havholm is on a break from her teaching job at the University of Wisconsin in Eau Claire.

As we sped along the nearly deserted highway, the geologic history of the Colorado Plateau revealed itself in pale slopes of Kaibab limestone, in greenish hills of the Chinle formation. The towering red cliffs of the Kayenta formation. The marbled red and white of Navajo sandstone, capped here and there with the aquatic sediments of the Carmel formation. In her mind Havholm holds an intricate map of the rock layers, their bends and plunges and skyward thrusts. "For some reason the formations on the Colorado Plateau have behaved as a solid block during compressional and extensional events," she said. "The sediment layers remain basically intact, one on top of the other, draping over ancient faults. So you get a lot of dramatic dips." As we drove, layers appeared and vanished, rising precipitously and diving beneath the earth's surface like steel girders twisted by an unfathomable force.

More than 300 million years ago, during the early part of the Paleozoic era, much of the American Southwest lay beneath a salty body of water. Over time the sea expanded and contracted like a pulsing cell, leaving its mark in extensive hardened sediments that scientists now use to delineate geologic events. Later, during the Pennsylvanian period, all the continents of the world were fused together, and the terrain that would become the Southwest lay well south of the equator. It was a hot, dry, wind-plagued landscape. Those areas not under water lay beneath sand. As the continent drifted north across the equator, the climate briefly changed. The air grew moist; plants flowered and dinosaurs roamed, laying down evidence of their lives in fossils and petrified wood. Northward pushed the land, back into the dryness of the horse latitudes. Once again small arms of ocean began to encroach from the west, a tide going in and out to a geologic rhythm.

We stayed the night in a motel on the south rim of Lake Powell. The following morning we were up at dawn to watch the spread of winter light across the rugged terrain. As we stood on the edge of town, our breath smoky in the cold, a small plane buzzed in the still air. Steam rose from the stack of a coal plant run by the Navajo Indian tribe. In Edward Abbey's book *The Monkey Wrench Gang*, eco-warriors conspire to blow up the railroad line that supplies this plant with coal, as well as the Glen Canyon Dam, which chokes the Colorado River to create Lake Powell.

To the east stood Tower Butte, a square-topped column of rock that appeared rosy in the slanted light. I asked Havholm if it had been sculpted by wind.

"My impression is that this particular area is mostly fluvial," she said. "It was shaped primarily by rivers and mass wasting — big slides and sediment movement caused by weathering and undercutting. Look across that mesa. Doesn't that look like an old streambed?"

Indeed, in the flat sweep of yellow soil I could trace the course of an ancient river.

We returned to the car and drove to a rise, where we could see vermilion cliffs dropping into the lake and, downstream from the dam, the jagged cleft of the river canyon. In the pure light of early morning, the earth seemed to have been dyed red. We followed a road south to a turnoff near a lopsided mound known to the geologists who study it as Julia's Knob.

The rock formations of the Lake Powell region include some of the most extensive units of aeolian sandstone in the world. Until the 1970s it was widely thought that these were formed by two ancient sand seas, the Navajo and the later Entrada. Then investigators identified a third rock unit, the Page sandstone, which was younger and thinner than the Navajo and separated from it by what geologists call an unconformity, a rock surface — in this case a surface overlain by wet sediment — that differs markedly from what lies below and above it. "The casual eye would never be able to see the difference between the Navajo and the Page," Havholm said. "You have to look real close. But the Navajo was basically a dry period. In the Page you can see big dune pulses, then a period of fluctuations — dry, wet, dry, wet — that's related to the expansion and contraction of the Carmel Seaway."

In the mid-1980s Gary Kocurek and another aeolian geologist, Ronald Blakey, agreed to try mapping distinct layers of Page sandstone throughout the Lake Powell region. It was a mammoth project. "No one had ever done that kind of detailed work in an aeolian sandstone unit before," Havholm said. No one knew the patterns of the dunes' movements or the wind regimes that had shaped them. Kocurek's idea was to dissect the Page sandstone to see if different events could be correlated from butte to butte, mountainside to mountainside. To complete the research he enlisted the help of several graduate students, including Karen Havholm.

We walked toward Julia's Knob on an uneven shelf of sandstone. In the middle the surface dropped in small, flaking steps, like a terraced garden. "Look here," Havholm said, pointing to an outcropping on the edge of the shelf, a wall eight feet high. On its side we could see three wide layers, as separate and distinct as those in a cake, streaked with faint red and ivory stripes. "Each one of those layers was put down by a different dune moving through," Havholm said. "Now step back a little. See how this red band comes across, and then the top of the sediment package is white?"

I did indeed. It looked as if a thin tongue of red rock had licked into the top layer from the side. Above it the rock was paler, though still predominantly red.

"That red rock is the bounding surface," Havholm said. "A time of no deposition in the dune field."

We continued on. The coolness of the desert air, with its tang of sage, was as pure and welcome as water from a spring. Havholm stopped walking abruptly and bent to the ground to show me a series of odd, gray, wishbone-shaped ridges a few inches high and as smooth and hard as bone. I knelt beside her. The ridges were interconnected; they continued across the ground for many yards in an almost geometric pattern.

"These are like giant mud cracks," Havholm said. "You don't get these in dry sand. This surface must have been damp and salt-encrusted sometime; it's the unconformity between the Page and the Navajo. It's how geologists first recognized the difference between the two formations."

We had almost reached the base of Julia's Knob, a lopsided dollop of Page sandstone. On its side I could pick out separate layers that delineated the remains of different dunes. Each had a thick base of steeply tilted stripes, topped with a lip of harder sediment, a bounding surface where wetness had intruded. In the stripes Havholm could trace the changeable nature of wind.

We began to climb, moving easily up the textured rock. Despite its name, slickrock is not necessarily slippery. After a few steps Havholm stopped me. "See these lines?"

Once again each layer was painted with steeply tilted streaks of red and white. Within the red stripes Havholm pointed out bumpy, wafer-thin lines: ancient wind ripples, hardened to stone. They would have been formed by light winds that moved sand across the toe of the dune. The smoother, wider white streaks showed where the winds had strengthened and carried much larger volumes of sand over the dune crest. "If these mark a seasonal wind shift, and I bet they do, you could use them to count the years it took the dune to migrate through," she said.

We climbed to the top, examining bounding surfaces, the polygonal fractures of mud cracks, the slouch of petrified dunes. In every direction stretched pinkish rock, crimped and cleft into odd figures. "The neat thing is, you can walk up that knob" — Havholm pointed north to another misshapen lump — "and find the same surfaces. Not one for one, but just about. Same with that knob over there." She pointed south. After the study, Kocurek and Julia Knight, the woman who mapped the geologic features of the knob, combined their findings with data collected by Havholm to reconstruct the wind regimes of this landscape hundreds of millions of years ago. They concluded that during the summer the Paleozoic winds had blown strongly from the north along the shore of the Carmel Seaway as the bottom third of Julia's Knob was being laid down. During this period the winter winds were a bit more variable, blowing predominantly from the northeast as a sea of large, compound, crescentic dunes moved across the land. A few million years later, after another intrusion of the seaway, the summer winds still came from the north. But the northeast winter winds may have become stronger, retarding the progress of the smaller, simpler dunes that were then migrating through. Kocurek and Knight could not tell how fast the wind blew or exactly how changeable it might have been. But through clues left in the rock, they managed to discern its major patterns of flow.

Below us, two ridges of flaky sandstone extended toward us like huge toes. "That's the Navajo formation," Havholm said. "That's exactly how those dunes were shaped as they moved across this plateau.

"To me, that's the important part. You can look at what's left and see how the dunes moved through. From that you can reconstruct what happened through time."

WE MADE our way back to the car and drove on. A short time later we were exploring the Ferry Swale mesa north of the Colorado River in southernmost Utah. Again the terrain was pinched and wrinkled; every speck of dirt was deep red. I walked down the sides of slickrock loaves with anvils of rock balanced on their backs. I climbed across stony waves topped with sharp formations, like bursts of spray. This is a hardened ocean, I thought, a rock version of the dunes beneath Nags Head Woods.

During her graduate work Havholm divided this mesa into six sections and walked each one, mapping every inch, drawing an internal picture of Page sandstone. Later she and Kocurek took their knowledge of ancient dunes and applied it to studies of modern dunes on the Texas coast; in White Sands, New Mexico; and in the African country of Mauritania. In each case they asked the same question: how do sand seas form and dissolve with changes in climate, wind, tectonic activity, and the presence of water?

Havholm had promised to show me as much variation in aeolian rock as possible, and she was true to her word. We drove washboard roads to an outcropping where a purplish line of volcanic ash lay exposed between layers of Page sandstone. We examined roadcuts to compare the coarse, pebbly sediments carried by water to the finely grained sand transported by wind. We studied the walls of the Entrada sandstone formation, laid down by changeable winds that left undulating layers swirling and crashing against each other, like current-pulled waves.

And we hiked a dry wash into a slot canyon with walls barely wide enough in places to allow passage of our shoulders. The slickrock walls tipped crazily inward, coming even closer together at the top, blocking the light. The air grew cool and moist. Above us, dust fell from the rim in fine cascades. Time stopped. We walked slowly along the smooth, sandy floor, at once enchanted and a bit on edge. Rounding a curve, we came on a huge root ball that had lodged between the walls, eight feet up, during a flood. "If it's raining anywhere upstream from here," Havholm said as we examined the unscalable walls, "we're dead." A minute later she ran her hand lovingly over the stone.

Now we are seated high on a marbled sandstone turret, being beaten up by wind. Grit has already seeped far into my mouth; during gusts I squint in an attempt to keep it out of my eyes. Silver, dry-stemmed bushes rattle like children's toys. Havholm sits beside me on the narrow ledge, balancing a sandwich on her lap, utterly at home. Our backs rest against a cliff of evenly layered Navajo sandstone ribbed with red and white stripes, wind's laminations and ripples. Below us dust dances a devilish whirl across a table of rock.

I close my eyes and see fiery rock set against a blue, blue sky. When I open them, I see a broader vista, columns and buttes and valleys of stone to a distant horizon. "To think this was once all sand," I say, "an incredible, never-ending sea of sand."

"Yes," Havholm says, "and we're sitting almost exactly in the middle of it."

I cannot absorb its vastness.

IN THE SUMMER of 1995, hoping to find a clue to the genesis of the dunes that lie beneath Nags Head Woods, Karen Havholm asked Glen Berger, a geologist from the University of Nevada at Reno, to date some sand samples from her study area. Berger was setting up a laboratory for measuring the luminescence held by sand that has been cored from deep within dunes. When a grain of sand is buried, it is bombarded by radioactive decay particles that give it a slight charge. Once it is exposed to sunlight again, the sun's energy drains the charge. Through a new technique, scientists can expose sand to light from lasers, then measure the intensity with which the grains luminesce. The greater the light given off, the longer they have spent in the darkness underground.

Havholm held great hopes for the tests to be run by Berger. Her study of the coastal dune system was still lacking a time scale; to draw any sound conclusions she badly needed to date the dunes. "Run Hill and Jockey's Ridge likely developed during the Little Ice Age [from the mid-sixteenth to the eighteenth century], because sea level would have dropped slightly then, and there would have been an increased supply of sand," she tells me. "But the dunes beneath the woods — who knows? They're probably less than five thousand years old. That's when the barrier islands would have formed in the vicinity of where they are now." Berger had promised to test the samples from Nags Head Woods as soon as possible. But that fall a key piece of equipment in his laboratory broke. Other setbacks soon followed. In the summer of 1997, Havholm is still waiting for dates on which to hang her conclusions.

And yet she believes she has solved the mystery of the odd assemblage of ridges that rise from the sandy earth like a set of interlocking waves.

One day I hike with Havholm back to the grove deep in the woods where she and Kris Weaver first dug into the intersection of a parabolic dune and a smaller, more westward-running ridge. Three long years have passed. Weaver has graduated; other field assistants have come and gone. In 1996 Havholm and two students dug a second sixteen-foot-deep trench into the side of this parabolic, still searching for a buried line of soil. At the bottom of the trench they used an auger to bore down to twenty feet. They found nothing but sand. Havholm also discovered that the strata from the parabolic exactly matched the strata from the small dune.

"It occurred to me, finally, that if I wasn't finding a soil line, there might be a reason for it," she says with a short laugh. "My thinking now is that these are all remnants of the same dunes. The smaller ridges are just parts that were left behind as the parabolics moved through." During periods when the water table was high, the base of the dunes would have been held in place by moisture. But the top portions, which were dry, would have migrated on.

This summer Havholm is working with ground-penetrating radar, a new technique in which radar signals reveal the location of underground anomalies, such as a water table or a buried line of soil. By using radar on Run Hill she has traced the location of an old layer of soil, the remnants of a vegetated dune. In some cases the radar can even reveal the stratification patterns within old dunes. "This technique is going to save a lot of trenching," she says. "We dug down where the old soil line was close to the surface and collected some material that we can date with carbon-14. It won't be real precise. But with that and the dates from the luminescent work, we should get a real solid idea of the time line."

With relentless determination she is teasing out the secrets of the dunes' past. It occurs to me that geologic studies require an abundance of patience and persistence, probably more than I could ever muster. "Do you think you'll ever get to the point where you're finished here?" I ask.

In answer Havholm grins at me, shrugs, and hikes on.

SEVEN

The Encircling Seas

ON AN UNSEASONABLY WARM March afternoon I stand on a concrete pier above a beautifully serene Atlantic Ocean, watching. There is a building west wind, but it has not yet begun to wrinkle the skin of the sea, at least not here, 1,800 feet out from an Outer Banks beach. Unbroken swells roll gently toward the pier, stretching upward and narrowing in girth as they "feel" the rising bottom. The muddy green surface of each wave is dappled by hundreds of wavelets that reflect the deep blue of the sky. These spangle the water with an ever-changing loveliness, as ephemeral and mesmerizing as the foaming of surf.

At the end of the pier is an instrument called a K vane, a gauge shaped like the letter K that measures the drag of the wind across the ocean surface. Beside me stands Charles Long, an oceanographic researcher in the U.S. Army Corps of Engineers, which owns the pier in the Outer Banks village of Duck and most of the instruments that dot it. To the northeast bob six large yellow buoys that anchor a field of underwater instruments. A bit south of the buoys the ocean surface is flat where a curious slick causes the wavelets to disappear from the faces of swells. Organic material has collected here, perhaps because of the action of internal waves.

The K vane points into the wind, the gray impellers on its arms whirring. Below the pier, waves roll to shore, coming mostly, as Long notes, from the east. "But now watch closely. Here comes a small one from the northeast . . . ," he pauses as it and another, larger swell from due east pass, "and now here comes one from the southeast." In the late 1980s, Long discovered that even waves formed by a wind blowing due east have a spread of about forty degrees. His findings have significant ramifications, since breakers approaching from various directions will erode shorelines differently from breakers that wash in from a single front. They also place an assortment of stresses on oceanfront buildings, jetties, and piers. "They're not unidirectional at all," he says. "They're incredibly complex."

It is the simplest and most telling statement anyone could make about ocean waves and the wind that drives them.

Yet for all its intricacies, the ways in which wind ripples the water surface is the best understood of its many effects on the oceans. Wind is both the sculptor of waves and the engine of currents. It is the spoon that stirs layers and helps create the flows that carry oxygen to the bottom and nutrients to the surface. Wind-driven currents sweep larval organisms away from deep ocean waters filled with predators and into coastal estuaries, where they can grow.

In polar regions winds and currents combine to keep the oceans unfrozen and to create the same kinds of drifting ice — smooth, rubbly, or hummocked — in the same places, year after year. Throughout the world wind carries the plumes of silt and toxins from rivers either offshore or spiraling back against the coastline, depending on its day-to-day whims.

And wind blurs the line between air and water, pulling vast quantities of heat and moisture into the atmosphere, changing patterns of weather around the globe.

TO UNDERSTAND how wind affects water and vice versa, one must remember that air and water are both fluids and subject to the same laws of physics. Air moving across water causes it to break into ripples, and then waves, that grow higher with time. Similar undulations form in moving air. At the sea's surface the two wave systems are coupled; they roll and push against each other. Eventually a building sea may actually decrease the speed of the surface wind by muddying its flow. This is the simplest example of what scientists call the ocean-atmospheric link, a continuous loop of feedback between heat and kinetic energy in the world's large waters and heat and kinetic energy in the air. In addition, in certain regions deep ocean storms of intense magnitude occur every two to three months. The storms send currents roaring across the bottom, mirroring the circulation patterns that push wind across land. It seems that there is not just one system of oceans on earth but two great encircling seas, one below the surface and one above.

The most obvious way that wind affects the oceans, besides making waves, is by driving surface currents. Water moves much more slowly than air — at about a hundredth of the speed of the winds that propel it — and it is deflected at a greater angle by the Coriolis effect. All wind-pushed water has a curl to its motion, to the right in the northern hemisphere and to the left in the southern hemisphere, because of the spinning of the earth.

At the latitudes just north and south of the equator, the flow of water is toward the west, driven by the trade winds. (Between the North Equatorial Current and the South Equatorial Current, in the zone of the doldrums, a counter-current flows east.) Where the equatorial currents bump against landmasses, they feed into other currents that follow continental coastlines. In the Atlantic, the North Equatorial Current merges with the Florida Current and then the Gulf Stream, one of the most powerful currents in the world. The Stream travels up the East Coast of North America and turns back toward Europe under the force of the prevailing westerlies, becoming the North Atlantic Drift. As it approaches Europe it turns south once again and merges with the Canary Current, which closes the loop. This great clockwise circulation pattern is called the North Atlantic gyre. Other gyres turn in the southern Atlantic, the northern and southern Pacific, and the Indian Ocean. They owe their existence not only to wind but to the arrangement of the continents. In the latitudes just north of Antarctica, where there are no landmasses to block the water's flow, gales drive an immense current called the West Wind Drift, which circumnavigates the globe.

While surface winds set up the great oceanic flow patterns, on a day-to-day basis they do not have much effect on the course of those patterns. As late as the 1960s ocean currents were assumed to be like rivers within the sea that kept to narrow, predictable paths. Thanks largely to satellite maps of sea-surface temperatures, we now know that the Gulf Stream, the Kuroshio Current off the coast of Japan, and other major currents form meandering loops that change over time; they wander, much the way a garden hose will writhe across a lawn when it is turned on full force. Scientists have found that the paths taken by major currents do not seem much affected by even major windstorms. Instead they are driven by physical differences within the ocean waters — by changes in temperature, density, and salinity.

Wind-driven currents serve as boundaries for marine plants and animals, many of which can live only within a narrow range of temperature and salinity. Except for mammals and birds, marine animals do not have the ability to regulate their body temperatures; nor do marine plants. As a result, for many species extreme variations in temperature can be lethal. In the deep ocean waters and in the midst of the gyres, water conditions change little, and life is fairly stable. But along coastlines and on the edges of major currents, temperatures may rise or fall abruptly as different water masses bump and squeeze against each other.

One region of great mixing is off the Outer Banks, where the Gulf Stream curls close to shore and a cold, southbound current, an offshoot of the Labrador Current, pushes against the Stream. The junction of these two flows forms a zone where marine organisms from both northern and southern latitudes can exist within easy proximity. The waters off Cape Hatteras are known for their rich mix of species, including game fish like tuna, dolphin, and marlin. They are known, as well, for their stormy nature. The sea-surface temperature can change by as much as 22 degrees Fahrenheit in only two nautical miles. This sudden shift plays havoc with surface winds.

In winter, when a mass of arctic air from the north hits the Gulf

Stream, it immediately begins to warm and rise. More air rushes in to fill the void left by the rising currents. At the surface of the sea the wind speed suddenly increases. The dramatic shifts in air temperature set up winds that are not smooth but highly gusty. And the collision of northerly (south-flowing) winds with the north-flowing current forms steep waves with short periods — that is, short distances from peak to peak. Waves that move so quickly can be extremely difficult for ships to ride out. This phenomenon is well known, and dreaded, among sailors.

To some degree meteorologists can adjust their marine forecasts to anticipate the elevated winds and seas that may build when an arctic air mass crosses the Gulf Stream. Even so, forecasters for the U.S. Navy in Norfolk, who are charged with safely routing the service's ships throughout the Atlantic, often underestimate the speed of southerly winds near the treacherous Diamond Shoals off Cape Hatteras, in summer as well as winter. "Something happens out there, some local phenomenon, that we're just missing," one meteorologist told me. "The wind speeds we predict will be off by ten or fifteen miles an hour, sometimes more. It's got to have something to do with temperature differences between the Gulf Stream and the cooler water along the coast. But in terms of sea conditions, it makes a huge difference."

WHEN WIND BEGINS to blow over an expanse of water, it scrapes the surface and forms the small bumps and ripples known as capillary waves. Gusty winds set up the dancing wavelets that mariners call cat's-paws, which enable sailors to see approaching wind as it moves across the water. These ephemeral mosaics allow the wind to gain more of a hold on the water's surface.

As the wind rises, it pushes up small waves in rhythmic patterns. On days when we want to go sailing, our family drives out to the banks of Roanoke Sound to look at the water so we can gauge the speed of the wind. With a breeze of ten miles an hour, waves march steadily across the sound; we rig our boat and set sail. By about sixteen miles an hour, the waves have grown so steep that they begin to break; small whitecaps flare briefly and fade away. We know sailing conditions will be fairly rough for our old day-sailer, but sometimes we launch for the challenge.

The size of the waves that a wind can build depends on its speed, its duration, and its fetch — the distance over which it blows. In the middle latitudes, where westerly winds prevail, the eastern half of an ocean tends to have rougher seas than the western half; because of the longer fetch, the wind is able to set up greater waves. The same is true for smaller bodies of water. On a summer day of stiff southwest wind, we restrict our sailing to Shallowbag Bay, on the east side of Roanoke Island, because we know the waves there will be no more than a foot high. Two miles farther east, on the back sides of the barrier islands, the wave heights may be two to three feet.

In terms of stress, wind acts on water as it does on animals and people and buildings: as its speed increases, the force it exerts rises exponentially. A thirty-mile-an-hour wind that blows across an unbroken expanse of ocean for a day and a night is capable of building seas of much greater magnitude than a twenty-mile-an-hour wind of the same fetch and duration. As a result, a small increase in wind speed can quickly change sea conditions from safe, if somewhat rough, to dangerous, especially for small vessels.

How a boat rides in rough waters depends not only on the height of waves but on their shape. Strong, gusty winds often build towering, whitecapped waves with short periods. The same amount of wind will build steep waves more quickly in shallow water than in open ocean, because the waves scrape against the bottom. In the waters of the Outer Banks, boaters know it is more dangerous to get caught in a storm in Albemarle or Pamlico Sound than in the ocean. The sounds are only six feet deep in most places; threatening waves can build on them in a few minutes of stiff wind. And although they are not as tall as ocean swells, the waves that form in the sounds have extremely short periods; each one hits a boat's hull within a few seconds of the one before it. Many local fishermen would rather ride out a nine-foot sea than take the relentless pounding of three-foot waves on the sounds.

Local watermen also know that windstorms can cause sudden and drastic changes in water depth. In shallow bays and sounds, hard winds push water in front of them so that it piles up against the far banks, exposing the bottom on the windward side. Several times a year a northeaster will push water to the western sides of Albemarle and Pamlico sounds, flooding marshes there but leaving bare sand on the sounds' east edges, just behind the barrier islands. If the wind suddenly slackens, the water is apt to rush back in, surprising anyone who has walked out on the normally submerged sand flats to look for oysters.

Wind-driven tides occur in many places around the world, not only in bays and sounds but in long, narrow lakes and seas where gales can develop a long fetch. A few scientists believe that a falling tide propelled by wind may have changed the course of history in the Gulf of Suez, just north of the Red Sea, more than 2,500 years ago.

It is a story known to every child raised in the Judeo-Christian faith. "Then Moses stretched out his hand over the sea; and the Lord drove the sea back by a strong east wind all night, and made the sea dry land, and the waters were divided." The Israelites, pursued by the Egyptians, walked into the sea on dry ground, "the waters being a wall to them on their right hand and on their left." Under Moses' guidance they made it safely to the far shore. But the Egyptians perished in great numbers when the divided sea collapsed back on them. The Hebrew name for the body of water crossed by the Israelites is Yam Suf, or Sea of Reeds, but no reeds or papyrus grow on the banks of the Red Sea. Biblical scholars now believe the crossing occurred not in the Red Sea proper, which is 2,000 to 3,000 feet deep, but in the shallow Gulf of Suez at its northwest end or even farther north, near Lake Manzala on the edge of the Mediterranean.

Could a wind really part a body of water so neatly and so briefly? In 1992 two oceanographers, Doron Nof of Florida State University and Nathan Paldor of Hebrew University of Jerusalem, devised a mathematical model in which they showed how a strong wind from the northwest might have enabled the fleeing Jews to safely walk across a ridge in the Gulf of Suez, a body of water about 220 miles long and between 12 and 19 miles wide. The gulf is rimmed on both sides by chains of tall mountains, which funnel the wind along its length. The prevailing winds in the region are from the northwest and the southeast. A wind blowing directly from the east would be unusual, because of the surrounding mountains, and it would not have much effect on the surface of the gulf. But *Ruwach kadim,* the Hebrew phrase used in the biblical story, may also be translated as northeasterly, southeasterly, or wind from the desert.

To create their model, Nof and Paldor used winds blowing from the northwest at about forty-five miles an hour. In ten to twelve hours, they found, enough water would be shunted southward to cause the depth of the Gulf to drop by eight feet on its north end, enough to expose any sandy bars or ridges. Storms of such length are highly unusual in the region. To blow at such speed for more than six hours, the wind would have to be fueled by two back-to-back storm systems, a meteorological fluke that might occur only once in several thousand years and would indeed appear to be miraculous.

In 1879 a scientist traveling in Egypt and Palestine noted the presence of an underwater ridge on the north end of the gulf. If a similar ridge existed at the time of the Exodus, Nof and Paldor write, the heavy winds would have left it exposed. The fleeing Israelites would have been given the impression that the waters had parted just for them. And if the wind suddenly subsided, the water would have rushed back to normal levels in as little as four minutes — quickly enough to trap scores of Egyptian soldiers and chariots.

No walls of water would have stood to the left and the right of the Jews' bridge to freedom. But the study makes it clear that the parting of waters by wind is not necessarily a farfetched idea. The scientists end one of their papers by writing, "Believers can find the presence and existence of God in the creation of the wind with its particular properties just as they find it in the establishment of a miracle. Some may even find our proposed mechanism to be a supportive argument for the original biblical description of this event."

*

JUST AS THE WIND can sweep aside waters in shallow bays and seas, in open ocean it can beget a horrible, heaving violence, breathing life into waves a hundred feet high and perhaps higher — so high that no instruments have ever survived to record their passage. And every so often the interaction of a weather system and a sea sets up an exchange of energy that spawns unstable atmospheric conditions like those over the Gulf Stream off Hatteras, but on a much larger scale, an exchange of energy that causes a low-pressure system to deepen violently. Such tempests occur with much more frequency in winter than in summer. Meteorologists have a special name for these storms: they call them bombs.

On March 12, 1993, a low-pressure core formed over the western Gulf of Mexico and intensified explosively as cold air traveled over the warm ocean water. A few hours later, when the center of the storm passed just south of Louisiana, wind speeds had reached ninety-eight miles an hour. Meteorologists named it the Storm of the Century. The system made landfall on the Florida panhandle and traveled up the eastern seaboard, sending the water from bays and sounds sluicing across barrier islands. Here on the Outer Banks, Pamlico Sound flooded houses more than half a mile from its banks. Many feet of snow fell in the northeast United States and Canada's Maritime Provinces. Property damage was estimated at $6 billion; when the system finally collapsed, the death toll stood at 270. Off Cape Hatteras and the coast of Nova Scotia, buoys recorded wave heights of fifty-two feet.

In the 1950s, when scientists first began analyzing weather bombs, it was believed that such systems developed relatively rarely. Twenty years later meteorologists at the Massachusetts Institute of Technology spent three winters tracking explosive storms throughout the northern hemisphere. They discovered that bombs form rarely over land but with surprising frequency at sea. The one place bombs tend to affect land is in the United States, where troughs of low pressure sometimes form just east of the Rocky Mountains, then drift southeast to the Gulf of Mexico, where the cool air collides with warm water and sets up severely unstable atmospheric conditions. The trough spins into a low; barometric pressure falls a millibar an hour or faster for a day and a night. This was the genesis of the Storm of the Century, and thanks to atmospheric modeling, forecasters predicted it a week in advance.

For the most part, the strongest winds and highest waves generated by bombs and other intense storms are felt far out at sea. Coastal towns in the East may be assaulted by heavy maritime winds, but they are spared the worst punishment because the systems drift out to sea on westerly winds, releasing their greatest violence in the middle of the Atlantic. A typical northeaster is the backwash from a strong regional cyclone, in which winds rotate counterclockwise around a low-pressure core. The strongest waves are generated in the southeast, or advancing, quadrant of the storm, where the westerly winds are heightened by the forward motion of the system.

One system that did bring its fury to bear on the East Coast was the Halloween Storm of 1991, which melded with a passing hurricane to create unusually strong easterly winds and heavy surf. A large winter storm, spinning counterclockwise off Nova Scotia, collided with the remains of Hurricane Grace, intensified, and drifted back toward the East Coast. Meteorologists are still not sure what forces came into play that allowed the two systems to merge. To the west of the combined storms was a core of high pressure, with air spinning in a clockwise direction. The high- and low-pressure systems bumped together over the ocean like two huge gears. Turning against each other, they sent air racing westward, toward land, over several hundred miles. The storm reached its peak on Halloween.

In twelve years on the Outer Banks I have never seen the ocean unleash such violence. The Halloween Storm is the gold standard in these parts; it is the storm that for the first time made me fear the ocean. At force 12 on the Beaufort wind scale, the sea is described as completely white with driving spray. That is how the ocean appeared not only during the storm but a full day after the wind had calmed. Walls of water heaved onto the beach, one after another, with no lull between them. Over Georges Bank, peak winds of more than seventy miles an hour were recorded. Sea heights reached more than fifty feet, and one wave measured nearly one hundred feet.

It is difficult to imagine what it would be like to try to ride out such seas in a boat. When caught in a storm, mariners usually "heave to" — that is, they steer into the waves and simply hold on, climbing up mountainous peaks and dropping into troughs.

The first time I journeyed offshore on a fishing boat, the seas stood at about five feet, a little rough but not unusually so. The wind was from the east. At 5:30 A.M. the captain, a commercial fisherman I'll call Will, took us through Hatteras Inlet. I was riding with him on the bridge. As we made open water, the boat fell out from under us, then rose on a crest and fell away again. My stomach flipped. I thought, "I'll never get through this." But I am not prone to motion sickness, and I made myself relax enough to survive the ninety minutes it took us to reach the Gulf Stream.

For most of the day we set rigs for bottom fish, hoisting lead weights over the gunwales of the boat and immediately reeling them back in with electric winches. As we brought in grouper, sea bass, and triggerfish, the wind calmed a little and the sea rocked us benevolently. I found I liked the motion of the waves and the inky color of the water. The wind-rippled crests, rising above the boat, reminded me of polished stone. I was glad I had come. On the way back the wind was at our stern, and the ride was fairly smooth. But as we entered Hatteras Inlet, Will misjudged the size of a breaker on the bar. As he ran through it, the boat lurched to the port side and tipped. I was thrown against a metal support; I hung onto it, looking straight down into the ocean, for the heart-stopping moments it took the boat to right itself. Although I have been offshore many times since, that experience cured me of ever wanting to be caught at sea in a storm.

"Fishermen say they can gauge how fast the wind is — and how worried they should be — by the sound it makes against the wire stays and outrigger cables," writes Sebastian Junger in *The Perfect Storm*. "A scream means the wind is around force 9 on the Beaufort Scale." That is, a strong gale, forty-one to forty-seven knots. I have been at sea in such wind, but not far from shore, and only with following swells that the boat took with no trouble. Still, the captain made haste in heading for port. "Force 10 is a shriek. Force 11 is a moan. Over Force 11 is something fishermen don't want to hear." Much later in the book Junger writes, "Anyone who has been through a severe storm at sea has, to one degree or another, almost died, and that fact will continue to alter them long after the winds have stopped blowing and the waves have died down." It's true. The handful of people I've met who have lived through storms at sea tend to talk about their experiences only when pressed, and then not in very descriptive terms, even years later, when one might expect the trauma of the event to have passed. "All I can say is that it was an immensely humbling experience," a lieutenant commander in the navy told me of riding out forty-foot seas on an aircraft carrier in the North Atlantic. It is as if something inside the survivors of storms, some cognitive function, shuts down whenever that memory breaks the surface.

Of course, people have different tolerances for rough seas, and any group of mariners will hold a range of opinions about what constitutes dangerous conditions. Gauging wind speed and wave heights is a tricky business, so much so that oceanographers tend to discount eyewitness reports unless they are backed up by instrumentation. Most ships at sea have anemometers, but in storms and high seas the wind flowing through a shipboard gauge tends to be distorted because of the surrounding wave crests and the shape of the vessel itself. Likewise, perceptions of wave heights can be exaggerated by how the ship takes the swells. The most accurate measurements are considered to be from moored buoys and, to a lesser degree, from satellites that measure sea state. Occasionally these reveal wave heights that scientists find highly surprising.

On August 26, 1995, an atmospheric wave of low pressure developed between the west coast of Africa and the Cape Verde Islands. Over the next five days the system organized into a tropical depression and reached hurricane strength. Meteorologists named it Luis. By most standards Luis was unremarkable. The maximum sustained winds at the sea surface were measured at 140 miles an hour, typical for hurricanes over open water. It never made landfall but stayed well off the eastern seaboard of the United States and finally fell apart in the North Atlantic. For roughly twenty-four hours, however, Luis traveled northeast at about thirty-five miles an hour, which happened to coincide with the forward speed of a certain group of waves.

Although the winds within hurricanes do give birth to stormy seas, the size of the waves is limited because the fetch is short; the spinning winds tend not to stay over any one area of water for very long. (The storm surge that precedes a hurricane is caused not only by strong winds but by the combination of the traveling wind field and a drop in atmospheric pressure that actually pulls up the sea level, forming a swell.) The most treacherous seas tend to be built by large regional weather systems, not by comparatively compact, quick-moving tropical storms. The exception to this rule occurs when a hurricane happens to travel forward at the same speed as a group of waves. In such cases it is as if the wind is blowing over an unlimited fetch. The waves continue to build and build until the storm either slows or passes them by.

On September 9, the luxury ocean liner *Queen Elizabeth II,* bound for New York and just east of Luis, encountered a wave ninety-five feet tall, larger than any the captain and crew had ever seen. The ship dropped into a hole, and then the wave broke over the decks, causing minor damage to the bridge. The ship's report of a rogue wave might have been viewed skeptically by oceanographers, but for the fact that Luis also traveled over several buoys south of Newfoundland maintained by the Canadian weather service. One buoy recorded winds of nearly seventy miles an hour and the passage of a wave ninety-eight feet tall — which was as tall as the instrument could measure. It is possible that the wave was well over one hundred feet. Whatever its true size, it was one of only two or three waves of such size ever recorded in the North Atlantic.

Data from other buoys on the northeast edge of Luis show that for the most part wave heights were no more than thirty to sixty feet (still seas of frightening size). The ninety-eight-foot behemoth was truly a rogue, perhaps one of the largest waves ever generated by wind.

Scientists believe that, while rare, waves of one hundred–plus feet are more likely to form in certain regions of the oceans. But no one knows for sure, because vast areas of the world's waters are without instrumentation. Satellite coverage of ocean states is spotty; coverage by buoys is even more hit-or-miss. Oceanographers have gathered the most detailed information along major shipping routes, where captains issue regular reports of unusual conditions. But in the roaring forties and fifties of the southern hemisphere, where winds blow at great speeds unblocked for thousands of miles, knowledge of wave size is particularly sparse.

Rogue, or episodic, waves build because of a combination of wind fetch, bottom topography, and currents, which can alter the shapes of waves. They seem to form most often near a continental shelf. Off the coast of southeast Africa, along a busy shipping route where tankers and container vessels ride the south-flowing Agulhas Current through the Mozambique Channel, numerous rogue waves have been reported since the 1940s. Some researchers have speculated that wave heights in that region may potentially reach 190 feet, but the theory is controversial and no waves that large have ever been seen. A hundred-foot wave would be difficult to ride out; a wave nearly twice that tall would almost certainly sink anything in its path. A tanker or container ship 1,000 feet long would shear in two if a rogue wave left a substantial portion of its hull unsupported. Or it might crumple beneath the force of hundreds of tons of water breaking across its deck. Ships at sea have been known to vanish mysteriously, and in some cases scientists suspect they were taken to the bottom by rogue waves.

Oceanographers have long known that rogue waves might take a variety of shapes. In the mid-1990s researchers issued an international call for observations of unusually large waves in the region of the Agulhas Current. Based on pictures and videos, there is speculation that the current may foster the generation of towering waves with unusually long crests and breaking tops, like surfing waves. If so, study of them might help captains learn how best to handle their ships if they encounter one.

Since the 1960s oceanographers have also honed models for "hindcasting" waves — that is, analyzing wave formation during past storms. Petroleum companies and other businesses that maintain platforms offshore or structures along storm-lashed coasts often hire consultants to predict the potential damage from what are called hundred-year waves, the largest significant waves that may develop within a century. In gauging significant wave height, scientists take an average of the heights of the tallest waves, excluding rogues; this gives a measure of the state of the sea over an entire region. Maximum wave height, on the other hand, measures only the largest waves, including any rogues.

Statistically speaking, there is little chance that any one ship or drilling platform will be hit by a wave of maximum height, and it is nearly impossible to predict rogue waves. Engineers tend to design structures to withstand the highest significant waves that they are likely to encounter in a hundred-year period. "We're getting close to having enough data to draw contours for hundred-year waves around the world," says Vincent Cardone, a leader in wave analysis who operates a private consulting firm in Cos Cob, Connecticut. "We already have enough data to predict hundred-year waves along coastlines and major ship routes. But in another five years we should be able to take the historical data we have, plus the more recent observations from satellites, and extrapolate them out to get hundred-year waves for the whole world."

While scientists may have a rough idea of the fury that could be unleashed by any ocean, anywhere on earth, their forecasts will not have great accuracy for many years, if ever. Historical records of ocean conditions go back only as far as the 1950s, because most of the observations taken during World War II and the years immediately following were lost. In addition, the computer models that predict sea states assume there will be no major climatic change over the next hundred years. That is a chancy premise. The encircling seas have proved themselves fully capable of confounding science, of baffling the men and women who try to predict their whims.

*

On a sunny summer day of light east breeze, I leave my desk (again) and head for the beach, where — just as I thought — beautiful blue-green waves loll against the shore. I drop my towel on the sand and dash into the clear, clear water, arms out as if to embrace it. A moment later I am racing back to shore, my legs afire from the stings of jellyfish.

I should have known. The warm surface waters raked against the shore by easterly winds contain an assortment of fauna, not all of which are pleasant swimming companions. The gorgeous, pulsating purple blobs known as sea nettles always show up in midsummer. So do the megalops — a larval stage — of various crab species, which look in the surf like dissected pieces of crab but bite mildly when touched. Of course, the denizens that travel shoreward with warmer water are not all to be avoided. Sometimes sprigs of sargasso weed appear, sheltering tiny crabs and shrimp and even mottled yellow-and-brown sargassum fish, their faces covered with fleshy tabs. Pompano school in the shallows like silver moons. Manta rays leap from the water within sight of the beach. On starry nights loggerhead turtles haul themselves ashore to lay eggs the size of Ping-Pong balls in shallow, sandy nests.

Just as terrestrial animals do, marine animals employ wind to help them navigate, hunt for food, and travel from point to point. The petite jellyfish known as the by-the-wind sailor moves passively on the breeze, which catches the translucent, purplish, sail-shaped membrane it carries on top. Often these animals travel in flotillas. I have seen by-the-wind sailors washed ashore by the hundreds on the beaches of the Pacific Northwest. The larger, more dangerous jellyfish known as the Portuguese man-of-war has a sail-shaped crest it can inflate or deflate, so it can choose whether to drift along at a 45-degree angle to the wind or simply float on currents. During the 1940s and '50s, studies of the man-of-war found that each one tends to bulge a bit to the left or to the right, controlling whether the organism travels to the left or the right of the wind. In the northern hemisphere, biologists found, most of the men-of-war that were collected leaned, and drifted, toward the left. This seems curious, given that the long lines of weed that form on the open ocean deflect to the *right* of the wind because of the Coriolis effect. The researchers speculated that by moving to the left of the prevailing wind, a man-of-war may be better able to pass between lines of weed and avoid getting its long, trailing tentacles entangled. In the southern hemisphere, where the Coriolis effect pushes seaweed to the left of the wind, the few men-of-war collected bulged mainly to the right of their sails.

The wind's shaping of waves apparently provides some marine animals with clues about the directions in which they should swim to find safety and food. In the late 1980s and early 1990s biologists at Florida Atlantic University and the University of North Carolina at Chapel Hill proposed a theory that loggerhead sea turtles find their way from the beaches where they hatch to strong ocean currents by sensing the earth's geomagnetic field. In essence, the researchers believed that the turtle's brain circuitry contains an internal compass that tells them what direction to take when they enter the water. As part of their tests, the researchers took a number of hatchlings off the Atlantic coast of Florida and tethered them to a floating buoy to see which way they would swim. They repeated the test several times. On most days the animals swam vigorously eastward, as if sensing the geomagnetic field. But on an unusually calm morning the turtles swam in random directions, as if thoroughly disoriented. The biologists watched in astonishment. After a few minutes, the wind began to blow from the east — and the turtles turned in unison and headed out to sea. In later tests the researchers found that young loggerheads consistently swam into wind-generated waves, even when doing so took them back toward land. In mid- to late summer, the season of turtle hatching, the winds off Florida tend to blow from the east in the early part of the day, when the turtle hatchlings would be striking out toward the Gulf Stream. The investigators noted that by swimming into waves, the turtles were likely to make their way offshore, regardless of whether the beach where they hatched faced due east. Once they had traveled a distance offshore, the hatchlings might use the earth's geomagnetic field to lead them to the Stream and the beginning of a several-year journey that scientists believe takes them completely around the North Atlantic gyre.

Seabirds in the far South Atlantic have been shown to hunt by following the pungent wind-borne scent created by feeding krill, the small, shrimplike creatures on which they prey. When krill come to the surface to graze on plankton, they release the chemical dimethyl sulfide, which smells something like rotting seaweed. In experiments off Antarctica, researchers from the University of California at Davis and the University of Washington found that prions and Wilson's storm petrels traveled in large numbers to slicks of dimethyl sulfide that the investigators created artificially. They speculated that such wind-borne scents may serve as navigation guides for the roaming seabirds known as tubenoses, a group that includes petrels, prions, shearwaters, and albatrosses.

Perhaps more than any other animal, albatrosses are creatures of wind. Ten of the world's thirteen species live on the open seas of the southern hemisphere. Because they depend so on wind, only rarely do any cross the doldrums and make their way into the northern hemisphere. The wandering albatross, the largest seabird in the world, has a wingspan of eleven and a half feet, but its wings are only nine inches wide. It is as aerodynamic as the most sophisticated human-made glider.

Four species of albatross, including the wandering, have evolved to dwell in the seas around Antarctica, the windiest continent on earth. They spend most of their lives flying without wing beats, soaring fifteen to twenty yards above the ocean surface, then dipping low to the waves and riding the wind back up. The ecologist David Campbell, who has made a number of voyages through Drake Passage, writes of black-browed albatrosses, "At first they seem as stiff-winged as a child's kite, but in fact their wings are wonderfully sensitive and dextrous. The wing *feels* the subtle eddies of wind along its length. Each wing bone and feather has a suite of muscles that can slightly shift its orientation and, in unison, adjust the wing's total geometry — the angle at which it is swept back, the degree at which it cups the wind, the total surface area — from second to second." Of the wandering albatross, Campbell adds, "They seem to be objects of the sky itself, weightless beings that embrace the wind and float on it with a relaxed elegance. They are paradigms of economy and aerodynamic efficiency."

Wind also affects the movements of fish in ways that anglers tend to understand, or at least anticipate, more than scientists. The fishing lore of the Outer Banks and other coastal regions is full of advice about what species will be biting on which winds. A friend of our family, a man who runs a trawler off the South Carolina coast, told me recently that the direction of the wind is absolutely key to whether he catches shrimp. During a northeast or east wind he never takes the trawler out because the shrimp will be buried deep in the mud.

"Why would they bury themselves in an east wind?" I asked.

He shook his head. "They just do. They disappear. Ask any of the old-timers around here and they'll tell you the same thing."

In these parts charter fishing boat captains believe a northeast blow is good for catching tuna, though in the worst gales few boats venture offshore. One spring, when a northeast wind was blowing twenty-five and the seas stood at ten feet, some men I know went fishing in the Gulf Stream, sentencing themselves to a day in a thrashing, violent world. Waves rose around them like peaks. As the boat dipped into troughs, sea creatures floated by at eye level. One man looked up to see a manta ray with a wingspan of ten feet flying through a wave over his head. The men hung on for hours, getting sick and a little scared, until suddenly all their lines began to sing under the weight of big fish. The ocean became a storm of brown backs and golden fins as the boat shot through a school of thousands of tuna. It was the largest congregation of fish the captain had ever seen. They managed to catch a couple of dozen — and in ten minutes the tuna were gone. The anglers were all a bit woozy when they came ashore that night, but they were elated just the same.

OFF THE NORTHERN COAST of California, the winds of fall and winter tend to blow weakly and variably, and clear surface water washes against the shore. Then in spring, in the space of a few days, the breezes shift strongly to the northeast as high pressure builds over the western Pacific. Surface water pours away from the coast, and water from the bottom rises to fill the void. This flow pattern, known as upwelling, brings vital nutrients to the top of the water column and provides a bonanza of food for coastal phytoplankton, and through them for other marine species.

The upwelling that occurs off California is a prime example of how wind helps stir the oceans, sending nutrients from the depths to the surface. (In regions where currents driven by wind converge with currents flowing from another direction, surface water tends to sink, taking precious oxygen to bottom-dwelling creatures.) The same upwelling phenomenon that occurs in northern California is found also on the North Carolina coast, although with less reliability. Northeast winds nudge clear surface waters against the beaches of the Outer Banks, while southwest winds push them offshore. On the hottest summer days, when the wind blows ceaselessly from the southwest, the surf temperature will suddenly dip from the seventies into the sixties as bottom water is sucked toward shore. Often the bottom water is not only cold but as muddy as a Mississippi stream. Earlier in the summer, northerly winds brought plumes of silt from rivers and the Chesapeake Bay to coastal North Carolina. Now upwelling is bringing the silt ashore.

Wind-driven flow patterns along shorelines are far more complex than patterns in the open ocean. Huge gaps exist in our knowledge of coastal processes, in terms of both fluid dynamics and in the effects of flow patterns on the fauna of the sea.

In the early 1980s oceanographers conducted intensive studies of upwelling off the coast just north of San Francisco. They were surprised to find strong current jets — fast, narrow flows — that extended to as much as 300 feet below the surface. The jets apparently form because of the area's steep bottom topography. They carry large volumes of cold coastal water offshore and foster exchange between the continental shelf region and the open ocean. Their discovery emphasized the complexity of wind-driven circulation patterns. It also pointed to the need for more study of flow dynamics to minimize problems caused by pollution, oil spills, and the vast overfishing of the world's coastal waters.

Knowledge of coastal processes has increased dramatically in recent decades, thanks in part to improvements in moored instruments and satellite imagery of the seas and in part to a new cooperation between different scientific disciplines. Beginning in the late 1970s ocean researchers who had previously focused on one narrow topic (such as the biology, chemistry, or physical processes of the oceans) began conducting interdisciplinary studies on large-scale events, like upwelling off the California coast or the wanderings of the huge eddies that occasionally pinch off from the Gulf Stream and send a storm of water from the Sargasso Sea spinning south along the East Coast. In the late summer and fall of 1994 nearly a hundred scientists gathered at the Corps of Engineers' research pier at Duck for an intensive investigation of the effects of seasonal wind patterns and storms on the coastal region — the mixing of the open ocean and the waters on the continental shelf, the energy in waves, the transport of sand to and from the beach, and the migration of larval creatures, especially species that are important to commercial fisheries. In the summer and fall of 1997 another large group of researchers gathered at the pier to focus on sand movement.

The Duck research pier is an oddity, sandwiched as it is between some of the most upscale housing developments on the Outer Banks. The half-mile-long, pristine beach that it occupies is one of the most studied sections of shoreline in the world; many oceanographers regard it as the international center for research into coastal processes. Scientific instruments stud the pier, including a unique Sensor Insertion System, a kind of boxy, two-armed control house that rolls the length of the pier on rails and can insert gauges into the water at various depths.

On the day I visit Charles Long at Duck, I find a strange-looking vehicle, half truck, half boat, parked in front of the building that houses the offices of corps scientists. With its large tires and wide wheelbase, the LARC (as researchers call it) can both roll through surf and float in deep water. At the edge of the beach is a towering tripod set on wheels, the facility's famous roving research vehicle, known as the CRAB. The legs of the CRAB support a platform thirty-five feet above the ground. Scientists employ the device to map the bottom of the surf zone — a difficult task — using a combination of water-depth measurements and a global positioning system and, for more detailed data, a side-scan sonar system that can detect sand ripples.

To the north and south of the office stand towers with video cameras that continuously film the surf zone, charting the action of waves and mapping changes in the offshore bars. "Coastal scientists have gotten pretty good at predicting how sediment will move on an evenly sloping beach," William Birkemeier, the chief of the facility, tells me. "But on beaches where there's a bar offshore, it's amazing how much we still don't understand.

"For instance, you would think that when you have a wave breaking across a sandbar, there would be a lot of current on the bar. But that's not true at all. The current forms in the trough in front of the bar. We don't know exactly why that happens. There are theories out there, but we don't have any consensus on it."

In 1994, when the scientists gathered to collect data on coastal processes, one of the researchers was Cheryl Ann Butman of the Woods Hole Oceanographic Institution, who studied the movement of larval animals in different wind regimes.

Chuck Long's description of Butman's work piques my curiosity, and after my visit to the pier I call her. "Wind direction, and upwelling specifically, is really important in terms of carrying larval animals offshore and onshore again," she tells me. In northern California researchers have found that during times of strong wind from the north and northeast — times of strong upwelling — few barnacles attach to coastal piers and boats. Apparently most of the larvae are swept offshore by currents. But the larvae are not simply lost to the deep ocean. Instead, they bump up against the front where colder midocean water meets the coastal water pushed westward. When the northerly breeze slackens, the surface water moves back to the coast, bringing a pulse of barnacle larvae with it. "We're interested in seeing whether the same kinds of processes affect commercial species on the East Coast," Butman says. Mature clams might release their young when waters just off the beach suddenly cool, signaling a wind shift to the south or southwest and the start of a period of upwelling.

The clam larvae studied by Butman grow in the water column offshore, then return to the near-shore bottom to metamorphose into shelled adults. "Is there something that enables them to ride currents back to shore at the right time?" she asks. "We don't know. The initial indications are that they ride back in passively, then, if they choose to stay on the bottom, they burrow in. The surf clam is very particular; it won't burrow into a substrate that's muddy. It wants sand." In time Butman hopes to be able to predict where commercially valuable clams will be found within a region, based on wind patterns and the circulation of coastal waters. "These populations are really patchy," she says. "It would be ideal if we could develop a prediction model to show where the most recruitment will occur, and when. That's the ultimate goal."

The wind spreads young marine creatures just as it spreads the seeds of plants, letting some fall on fertile ground and whisking others back to sea. And it seems that the wind serves as sower not only with the bivalves that live in the seabed but with at least some species of fish.

James Ingraham, Jr., is an oceanographer and fisheries scientist with the National Oceanographic and Atmospheric Administration (NOAA) office in Seattle. Ingraham has long tracked wind-driven currents in the eastern Pacific, using a computer model and some rather unorthodox data. In May 1990 a container vessel traversing the North Pacific about five hundred miles south of the Aleutian Islands encountered rough seas and lost overboard containers holding 80,000 Nike athletic shoes. The spill occurred in an area of weak flow, where the remnants of the Kuroshio Current push water lazily to the east. Most of the surface currents in the region are driven by prevailing westerlies and flow toward the North American continent.

Seven months after the spill, some of the shoes began washing ashore in Oregon, Washington, and western Canada. Oceanographer Curtis Ebbesmeyer of Seattle collected reports about where the shoes washed up and gave them to Ingraham, who then used the computer model to draw a map of the wind-driven surface currents that would have carried the Nikes ashore. Most of the shoes drifted to the east. The largest number washed up on Vancouver Island and the coast of British Columbia just to the north, an area of protected inlets and bays and an important nursery for larval fish.

A little less than two years later, another ship lost containers holding more than 28,000 plastic bathtub toys — beavers, frogs, turtles, and duckies. Then in December 1994 a vessel caught fire in the north Pacific and spilled nearly 39,000 pieces of ice hockey gear, including gloves, shin guards, and chest protectors. After both incidents Ingraham and Ebbesmeyer enlisted the help of beachcombers to chart the trajectories of the lost items.

The picture of the surface currents that emerged from the studies have helped Ingraham draw more precise models of how larval fish move from open waters to the western North American coast. Of particular interest are data on walleye pollock, an important commercial species. Pollock spawn in the Bering Sea on the edge of the Alaskan continental shelf at depths of six hundred feet or more, and their larvae drift up to surface waters. "The edge of the shelf is a very productive area," Ingraham says. "There are a lot of predators, so the larvae need to get to inshore waters as quickly as possible." The survival rate of young fish tends to be much higher in years with strong southwesterly winds. "You can correlate it really well," Ingraham says. "If you have a year with a lot of southwest wind, three years later the boats catch a lot of pollock." To hone his prediction model and apply it to other species, he is compiling a model that can correlate wind patterns of past years with the number of fish caught commercially, going back as far as 1901.

But if the prevailing westerlies of the middle latitudes push larval fish east, ever east, how do species off the Atlantic coast find their way into protected habitat where they can grow?

The answer, it seems, lies with the odd weather patterns set up by the Gulf Stream. Since the early 1980s a group of oceanographers, most of them based at North Carolina State University, have examined currents and the movements of larval fish through the ocean inlets that divide the islands of the Outer Banks. The five species studied — spot, menhaden, croaker, summer flounder, and southern flounder — account for 90 percent of the commercial catch in the estuaries of Albemarle and Pamlico sounds.

The fish leave coastal waters in late fall and make their way offshore to spawn, assisted by currents set up by prevailing west and southwest winds. From November through March, four or five regional cyclones a month form off Cape Hatteras and drift north, throwing northeast winds against the coast and carrying larvae back to shore on subsurface currents. Northeasters cause water to pile up against the ocean side of the barrier islands and inlets, but they drain water away from their western sides. As a result, an intense pressure gradient develops within the inlet channels. Water jets through Oregon Inlet, through Hatteras Inlet, through Ocracoke Inlet, carrying infant fish into the protected estuarine waters.

In the mid-1990s the researchers conducted detailed studies of wind-driven changes in currents and the intrusions of salty ocean water in two large estuaries just south of the Outer Banks, Bogue Sound and Back Sound. Both are drained by Beaufort Inlet, which — unlike the Outer Banks inlets — faces almost due south. When larvae enter Beaufort Inlet, current and salinity measurements show that southerly winds are likely to carry them into Bogue Sound, while westerlies take them into Back Sound more often. "It's fascinating. You can actually see a large difference in flow and salinity and the presence of larvae in different areas, all based on wind direction," says Leonard Pietrafesa, a professor at North Carolina State University and a principal investigator in the studies.

Although minute in size, larval fish are energetic swimmers, and marine biologists are still unsure of the extent to which they may position themselves to ride a torrent of water through an inlet. Perhaps a change in smell or in the temperature or turbidity of the water gives them clues that it is time to go and that this inlet, this exact place in the water column, is where they belong. No one knows. But, Pietrafesa notes, "the fish have developed a behavior that best fits the wind patterns of the region. The more we look at the system, the more we find that to be the case. Evolution is so commonsensical, it's amazing."

THE WIND, the wind. How subtly it leaves its mark on the many facets of the natural world.

The breeze is still lightly from the northeast as I hit the beach with a recent acquisition, a twelve-foot ocean kayak that will (if I am lucky) carry me through the surf, past the jellyfish, and into the clear, deep waters offshore. I pull on a life preserver and wade into the surf with the boat, careful to keep it behind me so it will not bang against my legs in the backwash, as it has done before. Kayaking in surf can be a bruising experience. I point the boat straight into the waves, gauge the size of a looming breaker, and lunge for the seat. The kayak tips crazily to the left and levels itself just as the foaming wave hits. I paddle hard and plow through. Salt water streams off my arms and drips into my eyes.

By the time the next wave hits, I am far enough out to get through it before it has fully broken. The bow of the boat tips skyward, then plunges down as it tops the crest, slapping hard against the back of the wave, throwing me forward. I gather my wits and paddle, moving as swiftly as I can beyond the breaking waves.

For a day of light wind, the ocean seems to have an unusual amount of spunk. Triangular jade crests march ceaselessly toward me as I paddle east. Over the last two days the wind has been from the southwest, and the ocean is still turbid. How far must I paddle before it will clear? I have always wondered just how far out the west wind pushes the surface water. Today I hope to find out.

I am within easy sight of land, and my kayak is fully capable of riding these swells. Still, the choppiness of the sea makes me slightly nervous. In an effort to relax I scan the sky for pelicans. None around. Last week I made my way into a pod of dolphins and a flock of diving seabirds, but today I am alone. I wonder what animals may be moving unseen below.

I paddle, thinking about wind, ocean, and weather. Fifty miles offshore the Gulf Stream loops north, its water temperatures reaching ninety to ninety-five degrees. Scientists now suspect that both the North Atlantic gyre and the more northerly subpolar gyre tend to swing between two temperature cycles, hotter and colder, each cycle lasting about twenty years. The warming and cooling of such vast amounts of water must have pronounced effects on the flow of high-altitude westerly winds, the engines of climatic change. And recently I have seen several articles about a new El Niño. Sometime last spring the trade winds over the equatorial Pacific suddenly weakened. Now the warm waters of the central ocean are pushing against the west coast of Peru. This El Niño shows signs of developing unusually early in the season; meteorologists fear it may be the most devastating in fifteen years. A glitch has already formed in global wind patterns. The Indian and Asian monsoons are expected to be weaker than normal. Next winter heavy rains will fall in the American Southwest, while the Northeast will be unusually warm. At least, that is what computer models predict. For all anyone knows, El Niño will dissipate unexpectedly, as it has before.

The boat glides through the waves. I stretch my arms and legs, arch my back, and sigh, finally at ease. The blades on my paddle leave small whirlpools and oily wrinkles on the ocean's surface. To the right I glimpse a shape moving in the water. I strain to see through the silt, but it is gone.

Several hundred yards ahead is a line of lighter green, perhaps the point where the water clears. I'm nearly a half-mile from the beach. Do I dare go that much farther? I paddle hard for several minutes, but the line appears no closer. It occurs to me that this is a fool's errand. The wind may have pushed clearer water miles from shore. I concentrate on my strokes, punching forward, pulling back, striving for exercise, for good form, but also trying to keep an eye on the depths below. Ahead I see a brown patch in the water, and paddle quickly toward it.

It is a school of cow-nose rays, maybe twenty of them, swimming just below the surface. They sense my presence and bolt, but I stay with them for a few seconds, watching their squarish, mottled brown bodies fly into the silt-laden depths. The waters close around them. I sit still for a few minutes, letting the waves rock the boat, until the wind sets goose bumps on my damp arms and neck, and then I turn for home.

EIGHT

Of Body and Mind

IT IS LATE JUNE and ridiculously hot, nauseatingly hot, maddeningly hot. June is normally one of my favorite months on the Outer Banks, but not this year. An ill wind blows from the southwest, the blast-furnace breeze of full summer. The air is wringing wet, the sky an ugly yellow. Air conditioners moan boorishly beside every house. The ocean has lost its early summer clarity and turned sulky with silt; on the beach, slathered tourists struggle to erect barricades against blowing sand.

I don't want to touch anyone, or be touched. The irritability of our normally cheerful son annoys me. I have to look deep into the recesses of my mind to find anything approximating a glimmer of hope or joy. Exercising is out of the question; I long to retreat to a hammock with a cool drink.

This is the season when wind haunts me more than at any other time. In summer, wind's direction dictates everything: the feel of the day, the presence or absence of biting flies in our yard, the amount of energy coursing through my blood. When the islands are deep in the teeth of a southwester like this one, it seems that the oppressive heat will never break. Ah, but with the arrival of a new weather system, with a shift of breeze even briefly to the northeast, the whole landscape changes. The air becomes as cool and refreshing as that beside a mountain stream. People emerge from their indoor hideouts like refugees arriving home. I pull out my moldering running shoes and go for a long jog, rejoicing in the breeze on my face.

Throughout the world there are winds known for their poison and those known for their powers to heal. Wind can imbue us with physical vigor and clarity of mind or bring on an intractable muddleheadedness and a thirst that cannot be slaked. It can make us jumpy as cats one day and unflappable the next. Winds are said to affect the rate of heart attacks and suicides in a community, the incidence of crime. And these are merely the winds we encounter from day to day. The scars left by a catastrophic windstorm, a gale or hurricane or tornado, may alter one's entire life.

How does the passage of air affect the human physique, the human spirit? It is difficult to say with precision. Medical climatology is still an inexact science, in which an array of factors — heat, humidity, solar exposure, the presence of pollution, and so on — must be considered together. And so the answers to that question are nearly as numerous as the names for wind, and as complex as patterns of atmospheric flow.

NEAR THE TURN of the fourth century B.C., Hippocrates described the physical impacts of certain weather and winds in the book *Airs, Waters, Places.* "Whoever wishes to pursue properly the science of medicine," he begins, "must proceed thus. First he ought to consider what effects each season of the year can produce. . . . The next point is the hot winds and the cold, especially those that are universal, but also those that are peculiar to each particular region." Elsewhere Hippocrates blames the south wind for headaches, sluggishness, and dullness of hearing and vision, and the north wind for coughs, throat infections, constipation, and more.

During the centuries that followed, many others, from Aristotle to Saint Hildegard to Voltaire, wrote of the healthful and unhealthful qualities of various winds. Regional breezes and gales came to be known by folk names that signified their leverage over the people within their spheres. There was the bora of the Dalmatian coast, reminiscent of the cutting north wind that ancient Grecians called

Boreas. The harmattan or "doctor," a northeast trade wind of the Sahara famous for its hot breath and heavy burden of dust, was welcomed by colonists on the Guinea coast as a change from the damp marine wind. The crop-ripening *trauben-kocher* or "grape cooker" of Switzerland. The *shawondasee* or southerly "lazy wind" known to the Algonquin Indians of eastern North America. The samiel of Turkey and the simoom of North Africa, both from words meaning poison. The horrible khamsin of Egypt, taken from the Arabic word meaning fifty, because it blows for about fifty days. The "bad-tempered" *melteme*, a northeasterly of the Aegean Sea. The whirling *yamo* or "wind in the body" of Uganda. The downslope mistral or "masterly" wind of the Rhone Valley. The violent, wrenching *mezzar-ifoullousen*, a Moroccan southeasterly known in Berber as "that which plucks the fowls." In *Heaven's Breath*, Lyall Watson lists more than four hundred names for winds around the globe, many of them either laudatory or pejorative.

Physicians now know that the human body is encased in a thin layer of air that acts as insulation but that can be penetrated by wind. In a calm atmosphere this layer varies in thickness from four to eight millimeters — a sixth to a third of an inch — but it quickly thins in even a light breeze. Wind acts on humans in two ways: it alters the exchange of heat between the body and the surrounding air, and it increases evaporation from exposed skin, cooling it. Because of our complex systems of thermoregulation, winds that are cold and dry are more uncomfortable than those that are cold and wet; and humid, sultry winds, like the southwesterlies of the Outer Banks, cause more physical distress than hot, arid desert winds.

In 1939 Paul Siple, an Antarctic explorer, first proposed the creation of an index that would couple ambient temperature and wind speed to create a new measure of temperature known as wind chill. Siple and a colleague published a formula for calculating wind chill in 1945, but the National Weather Service did not begin using it widely until the mid-1970s. By then Robert Steadman of Texas Technological College in Lubbock had devised another, more accurate heat-loss index that includes not only wind but the intensity of sunlight, the amount of clothing worn, and the activity in which the person is involved. The Steadman index is more moderate than the Siple; on a sunny day with a temperature of twenty degrees and a twenty-mile-an-hour wind, the Steadman index might place the absolute temperature at five degrees above zero, while the Siple wind-chill index puts it near ten below. Nonetheless, the Siple index has continued to reign supreme among the public, largely because of its simple concept and catchy name.

Since the velocity of wind rises with height, not all parts of the human body feel it in equal measure. The same wind brings roughly double the stimulation to exposed skin on the face as on the knees. In addition, the pressure of wind increases exponentially with speed. When a breeze of ten miles an hour picks up to twenty, it exerts not just twice as much stress on the body (or anything else it hits) but four times as much.

How a person reacts to wind depends in part on his or her constitution. Moving air massages cutaneous blood vessels, and this sensation affects people differently. Most find that light wind titillates their skin, while strong, gusty wind pummels it unpleasantly. At speeds below seven miles an hour a person can barely feel the wind; above fifteen miles an hour it begins to become irritating. The body has the capacity to work up a tolerance, even a taste, for wind, just as it can work up more and more stamina for exercise. But an appetite for wind requires a hardiness not common among people who spend most of their lives indoors. When I first moved to the Outer Banks, the constant wind greatly exacerbated my feelings of loneliness and vulnerability. Even now I view the wind as an adversary when I am sick or in low spirits. On good days, though, I find that I crave it like a sailor. When I travel inland to less breezy climes, I feel stifled and half asleep.

Elderly people are often troubled by wind. In his book *Weathering*, Steven Rosen writes of finding an older woman clinging to a tree one afternoon in a strong winter wind. "I asked, 'May I help you? Is something wrong?' She said, 'Yes. I'm frightened. The wind and the cold. I have a heart condition and I can't breathe in the cold wind.' I placed my arms under hers and she clutched me like a drowning swimmer." The woman had reason to be fearful. An analysis of weather-related mortality across the United States shows that elderly people are more stressed by severe weather events, including wind, than any other age group.

Schoolchildren, on the other hand, seem to react to gusty winds with energetic play. "Wind always stirs up my children," says a woman I met recently, a longtime schoolteacher. "You know if it's gusty you're going to have a harder time keeping things under control on the playground. Unless it starts blowing too hard. Then the children get upset and want to come inside." Wind may agitate children even in the classroom. A 1990 study of preschoolers' behavior showed that during unstable weather and high winds children tended to seek out adults and their classmates instead of playing alone or working quietly at their desks. Joan Didion writes that during the hot California wind known as the Santa Ana, some teachers abandon all pretense of holding classes because students become too unmanageable.

It is difficult to separate physical reactions to wind from emotional ones, since our physical condition is so strongly interlaced with our mental well-being. A summer day at the seashore with a light ocean breeze makes most people feel renewed, in part because of the pleasant scenery, the warmth of the sun, the sensation of water on one's skin, and the rhythmic sound of ocean waves. All of these are psychologically soothing, but there is a medical benefit as well. The moist, clean aerosols whipped up by wind help open the bronchial tubes and are therapeutic for people with respiratory difficulties. Indeed, they make breathing easier for us all.

In a study of weather and mortality rates conducted region by region in the United States, Laurence Kalkstein of the University of Delaware and Robert Davis of the University of Virginia note that scholars disagree vehemently over the ways in which weather may affect human health. Teasing out one variable of weather for study is difficult at best. But in analyzing statistics for forty-eight cities, Kalkstein and Davis found that people were more likely to die when unusually hot weather occurred early in the summer and when unusually cold weather came early in the winter. Once people had become acclimated to the season, they became tougher. The duration of the bad weather seemed to matter more than the extremity of temperature. If a hot or cold snap continued for many days, it usually claimed more lives than shorter, more intense periods of similar weather. This corresponds to what psychologists have observed about stress during windstorms: the longer the wind continues to howl, the greater the mental anguish.

Kalkstein and Davis found that in summer, mortality was highest during periods of light wind. When extreme heat was accompanied by strong winds, fewer people died. This countered an earlier study by Robert Steadman in which high winds seemed to increase the physical distress caused by heat. It is difficult to imagine that any weather condition could be more wearing than the staggering humidity and scorching winds of the Outer Banks during a midsummer southwester. The nagging presence of the wind is like a rash that will not stop itching; it leaves one without a moment's peace. Nonetheless, urban centers like those from which Kalkstein and Davis drew their data are natural ovens; their paved surfaces, scarce greenery, and high pollution may increase the ambient temperature by ten degrees. I would not want to be stranded in a large city center during a spell of windless heat.

A Polish study published in 1961 concluded that winds of from nine to nineteen miles an hour caused an increase in blood pressure and in the incidence of heart attacks and strokes. Chronic pain, too, may intensify on windy days for reasons that defy precise medical explanation. Perhaps it is simply that wind causes some people unusual stress. A study of Chicago-area residents with arthritis found that the patients' pain intensified on windy days, even among those who did not believe that weather could significantly affect their illness. Wind in urban areas tends to be unusually turbulent because of its passage among many buildings; the investigators speculated that the extra gustiness may have been more difficult for patients to tolerate than the smoother gales of open country. The same study found that arthritis sufferers in rural North Dakota did not experience more pain on windy days. Either they were more accustomed to spending time outside in the wind or the flow was less bothersome.

Many doctors suspect that attacks of asthma rise in some weather conditions, including high humidity and light wind, but the findings of researchers have been contradictory. In certain circumstances, however, wind clearly contributes to respiratory distress. A Barcelona study found that on scattered days of easterly winds, hospital admissions for asthma rose dramatically. Scientists correlated the attacks with the arrival of barges carrying soybeans; wind off the ocean picked up high concentrations of dust and spores from the soybeans and swept them into the city. In California's Central Valley, residents talk of a respiratory ailment called valley fever, which makes its appearance in late spring and summer when fields are being disked and the wind picks up massive amounts of dust.

Most studies on weather examine the influence of winds that roll heavily across the land. Light breezes, of course, are universally beloved for their powers to cool, cleanse, and heal. At a library one day, tired of reading medical studies on weather, I took a break to browse through a popular women's magazine. Halfway down the contents page a headline caught my eye: "How to Feel Sexy at 10 P.M." Everyone needs a little advice, I thought, flipping to the article. "Consider creating a breeze in your bedroom," the author advised. And why not? Moving air makes us aware of our bodies. It cools the skin and increases not only the flow of blood to the surface but sexual desire. When I went home that evening, I made sure to turn on the ceiling fan in our room.

TO ME it seems that wind wields its greatest influence not in the realm of the body but in that of the mind.

In the late 1980s a team of psychologists in Edmonton, Alberta, treated a ten-year-old girl for a severe phobia of the wind. Even watching a windstorm on television would trigger anxiety in the child. "A tornado had gone through the area," says Don Massey, a psychologist on the team, "and she became quite traumatized, even though the storm hadn't come close to her house. The littlest gust of wind would bother her. She may have had some sort of anxiety disorder before, but now it developed into a full-blown phobia."

The psychologists began working with the patient through hypnosis and counseling. As part of her therapy, the girl was asked to make drawings of herself standing outside. At first, Massey says, all the trees she drew were dead. "They were classic traumatic drawings. After she got a little better she'd draw trees bent over by the wind, or trees with half their leaves blown off." In the end, team members concluded that the girl's fears were sparked not by the tornado but by the separation of her parents and by problems in school. After a few months of counseling, her phobia disappeared.

Massey has since come across a number of cases involving wind phobia. I asked him about a Dutch case I had read about, in which a woman became fearful of wind while visiting the grave of her mother, who had recently died. Over time the wind became a constant reminder to the woman of her loss. "I don't know that case," Massey says, "but it makes sense. There has to be a personal trauma that serves as a catalyst for the wind phobia. Wind is a powerful suggester. That's why it's used so effectively in horror movies. It triggers a lot of memories, especially from childhood."

Fears of natural phenomena, including wind, marked the lives of many European settlers who fanned out across the world's frontiers in the eighteenth and nineteenth centuries. From the Russian steppes to the mountains of Patagonia, the writings of homesteaders detail the mental hardships levied by the gales of unfamiliar lands. On the plains of North America, women especially seemed to suffer in times of high wind. In her book *Pioneer Women*, Joanna Stratton portrays wind as one of the most formidable adversaries faced by the settlers of Kansas. During an eighteen-month drought in 1859 and 1860, Stratton writes, blasting southerly winds killed both the crops and the hardy prairie grasses. Many prairie poets, too, filled their work with images of wind. "O, wind, you are hellish hot; death is the song you / sing," wrote C. L. Edson in a 1914 poem entitled

"The Prairie Pioneers." But while the south winds parched the settlers' skins and throats and frayed their nerves, they were infinitely kinder than the north winds of winter, when venturing outside meant risking one's life.

In 1925 the novelist Dorothy Scarborough published a book called *The Wind,* set in the late 1800s. (A silent movie was later made of the tale, starring Lillian Gish and — unlike the book — featuring a happy ending.) The story is of a pampered young woman who moves to the desolate plains of west Texas and is driven crazy by blue northers and southerly gales. "The wind was the cause of it all," the book begins. "The sand, too, had a share in it, and human beings were involved, but the wind was the primal force, and but for it the whole series of events would not have happened."

The protagonist, eighteen-year-old Letty Mason, arrives in Texas at the height of a drought. Raised in Virginia, she has been impoverished by the death of her mother and seeks refuge at a cousin's ranch.

On the train west, Letty is warned of the wind's power by a handsome stranger. "It's ruination to a woman's looks and nerves pretty often," the stranger says. "It dries up her skin till it gets brown and tough as leather. It near 'bout puts her eyes out with the sand it blows in 'em all day. It gets on her nerves with its constant blowing — makes her jumpy and irritable."

Letty dismisses the stranger's warnings with a casual wave of her hand. How horrible could a little wind be? But as soon as she settles into her cousin's house, the wind begins to gnaw at her with its constant sounds, its inescapable breath. She begins to think of it as a wild black stallion, "with mane a-stream, and hooves of fire, speeding across the trackless plains, deathless, defiant!"

A few months after her arrival, Letty is driven from her cousin's house by his jealous wife and forced into a loveless marriage. Slowly the wind becomes a demon bent on destroying her. During its rare lulls Letty grows convinced that it is only trying to trick her into coming outside where it can catch her and swirl her away. "Heaven was a place where there was no wind or sand. Hell was where there was nothing else!" Her youthful beauty vanishes in nine short months. When the drought worsens and her husband's small herd of cattle dies, she demands that he borrow money so she can return to Virginia. Her husband rides off in a huff, leaving Letty to face a gathering norther alone.

As darkness falls, Letty feels as if the wind is peering in the windows, ogling her, taunting her. It shakes the flimsy shack where she lives and sends sand catapulting through cracks in the walls. She is wretched with fear by the time someone comes to rescue her. It is not her husband but the stranger from the train.

Beside herself, Letty lets the stranger seduce her. The next morning, when he tries to force her to come away with him, she murders him. She buries the body in a sandbank, thinking no one will ever know of her crime. But the crafty wind betrays her. Springing up anew, this time from the east, it blows the sand completely off the body. Letty peeks out her window, sees the corpse, and surrenders to her fate. The book ends with these words:

"With a laugh that strangled on a scream, the woman sped to the door, flung it open, and rushed out. She fled across the prairies like a leaf blown in a gale, borne along in the force of the wind that was at last to have its way with her."

ON CERTAIN DAYS in late fall and winter, when a cold front pushes over the Rocky Mountains and the polar jet stream swoops low in latitude, a river of cold air crests mountain peaks and begins a precipitous downward slide. As it descends, gravity speeds its flow, creating a Niagara of air and a visceral hell. By the time it reaches the foothills the stream of air is tumbling fast enough to snap limbs from trees, explode fences, and vibrate buildings. Desiccating and unnerving, the downslope wind borrows a name from its milder northerly cousin: chinook, the eater of snow.

John Calderazzo and SueEllen Campbell live in the foothills of the Rockies northwest of Fort Collins, Colorado. In spring the rising hillsides west of their house are covered with dried grasses that lie flat like the fur of a great beast. All point downslope, as if they have been curried by months of wind.

In Calderazzo and Campbell's bedroom is a window that wraps around the southwest corner of the house; against it they have pushed the head of their bed. A blue spruce stands guard just outside. It is an attractive place to sleep most nights. But when a severe downslope wind begins, blowing incessantly at fifty, sixty, even eighty miles an hour, Calderazzo finds himself beset by insomnia. Blood races through his head; his mind whirls all night. He listens to the branches of the spruce scrape the window. The previous owner of the house planted it there after wind lobbed a picnic table through the glass. "No matter how many pillows I burrow under, I can hear our big cottonwoods creaking and cracking, waving their limbs like dervishes," he writes. "I'm desperate for sleep, desperate not to wake SueEllen."

Traditionally chinooks were the warm, comparatively light winds of the northern plains, blowing across the gently sloping terrain of Montana at, say, twenty-five miles an hour. In recent decades, though, the name has become popularly synonymous with the devil gales of Colorado and Wyoming that bring out the jumpiness in a person's nature. These are the winds that prod young men in mountain college towns to walk their girlfriends home at night, because, as one told me, "Everyone knows the rape rate goes way up during chinooks." They are the winds of violence, of heart attacks, of suicides. They are the winds of madness. In Wyoming a law dating from the mid-nineteenth century provided that wind-induced insanity could be a viable defense for murder. "You can hear chinooks coming way before they reach you," one woman told me, "and you know you're in trouble. They always come in the middle of the night. They roar and let up, roar and let up, and they don't stop for days."

"At the end of two days we had the impression that the wind was not a natural phenomenon but a personal affront aimed by someone at us, and us alone," writes Gabriel García Márquez in the story "Tratamontana." "At lunch on Tuesday [the porter] regaled us with rabbit and snails. . . . It was a party in the midst of horror." The wind of which García Márquez writes is a downslope or katabatic wind of the Mediterranean, a land gale from Africa, but cold, unlike the chinook. Before the episode ends, the porter commits suicide.

Chinook, foehn, Santa Ana, *koembang*, zonda. These and a dozen other local katabatic winds are like venomous brothers, usually hot but always dry and chafing to the nerves. They share origins, screaming down dry mountain slopes to assault foothills settlements and plague inhabitants with insomnia, uneasiness, and sometimes rage. Visitors to regions beset by such foehn-type winds (named after the famous gales of the Alps) often find them pleasant. To many longtime residents, though, it feels as if a dark cloud has enshrouded them. The poison winds, such as the sirocco of the Mediterranean, the brickfielder of Australia, the *xlokk* of Malta, the simoom of North Africa, and the khamsin of Egypt and Israel, all drain the energy from one's bones. They impart a sense of unreality, as if nothing one does matters much.

One January several years ago I happened to be visiting the mountains outside Los Angeles when a Santa Ana blew up. For the next several days I suffered an uneasy buzzing in my chest. I was on a condor sanctuary behind locked gates in country as wild as any in the Southwest, yet I continually felt a need to look over my shoulder, as if for gunmen or thieves. Several times I sat down in a sheltered nook to breathe and take in vistas of the desert hillsides. Despite their parched, spare beauty, despite the solitude of the refuge, I could not make myself relax. On the morning the Santa Ana ceased, I felt my mind clear. The air was sweet and ringingly cool. But my body ached, as if I had been in a traffic accident. As if the mugger I feared had indeed roughed me up.

"There is something uneasy in the Los Angeles air this afternoon, some unnatural stillness, some tension," writes Joan Didion in her essay "Los Angeles Notebook." "What it means is that tonight a Santa Ana will begin to blow, a hot wind from the northeast whining down through the Cajon and San Gorgonio Passes, blowing up sandstorms out along Route 66, drying the hills and the nerves to the flash point." And later, "It is three o'clock on a Sunday afternoon and 105 degrees and the air so thick with smog that the dusty palm trees loom up with a sudden and rather attractive mystery." Not thinking clearly, Didion goes to a nearby grocery wearing only an old bikini. A "large woman in a cotton muumuu jams her cart into mine at the butcher counter. '*What a thing to wear to the market,*' she says in a loud but strangled voice. . . . She follows me all over the store, to the Junior Foods, to the Dairy Products, to the Mexican Delicacies, jamming my cart whenever she can."

On nights when the Santa Ana blows, Raymond Chandler writes, "every booze party ends in a fight. Meek little wives feel the edge of the carving knife and study their husbands' necks."

In the mid-1970s an Israeli pharmacologist named Felix Sulman began testing the urine of people who were particularly bothered by the blowing of the *sharav,* another katabatic wind. Over time Sulman found that people who suffered from migraines, irritability, sleeplessness, and nausea just before and during the *sharav* all exhibited a tenfold increase in the amount of the stress-induced hormone serotonin. (While serotonin is generally associated with feelings of well-being, too much of it can cause behavioral changes. Recent research suggests that large amounts of serotonin may cause people to act in extremely aggressive, violent ways.) The patients in a second group were overcome with fatigue, apathy, depression, and hypertension; all showed a drop in their secretion of epinephrine and norepinephrine. A third, smaller group exhibited a mixture of the two syndromes. Sulman suggested that all the patients greatly increase their intake of fluids, potassium, and salt during downslope winds to compensate for the loss of moisture and minerals. He devised a treatment for each of the groups using a combination of drugs, including several forerunners of modern antidepressants that modify serotonin production. Then Sulman recommended an auxiliary treatment based on research of questionable merit. He advised the *sharav* sufferers to invest in air ionization machines, which, he said, would correct an odd atmospheric imbalance that seemed to accompany the poisonous wind.

Any atmosphere contains a number of charged particles, or ions, that have broken away from stable atoms. Places that are generally considered healthful and peaceful — mineral springs, the seashore, the bases of waterfalls — have an abundance of negative ions. These bind to particles of dust and pollutants and so act as natural air filters. In still, heavy atmospheres and rooms made stuffy by crowds, positive ions tend to predominate. During the 1950s and 1960s a smattering of studies provided some evidence that negative ions in the atmosphere lower blood pressure, aid in the healing of burns, and ease breathing in people with respiratory ailments. Positive ions were found to cause skin temperature to rise and breathing to become more labored.

Sulman's attempt to connect ill winds with positive ions might have been better received if he had been more careful of his facts. He insisted that the air both before and during a *sharav* was engorged with positive ions, a statement not supported by adequate research. Moreover, Sulman asserted without testing patients that the *sharav* and the foehn of the Alpines caused identical symptoms in the people who found them unpleasant. The *sharav* and the foehn are spawned by different meteorological conditions in different climates. Sulman's work was subsequently discredited.

Air ionization has since fallen out of fashion as a panacea for the treatment of various woes. To my knowledge no definitive studies have ever been done on the odd electrical charges that may accompany ill winds. If there is a physical cause for the madness created by pounding streams of air, it remains locked in the complex circuitry of the human brain.

THE SHAPE and intensity of a poison wind ultimately depends on the terrain over which it passes. In Montana, where the Rockies subside along a gradual slope that extends across half the state, katabatic winds travel downward gently for hundreds of miles before rolling themselves out. But in Colorado, where the mountains crash headlong into the plains, gales ride down a steep chute that concentrates their power within a few dozen miles.

West of Boulder, air crests the Continental Divide and begins coasting downhill, only to run into the jutting peaks known as the Flatirons, which toss the air skyward once more. During strong downslope winds a standing wave of air may form over the Flatirons, much like the breaker that froths over an ocean sandbar in a strong current or the riffle caused by a boulder in a rushing stream.

When meteorologist John Weaver moved to Fort Collins in 1980 to take a job at Colorado State University's Cooperative Institute for Research in the Atmosphere, little was known about the conditions that combine to form severe downslope winds in most regions of the United States. Meteorologists had carefully studied the flow of wind in the volatile Boulder area, but no one had refined a technique for predicting downslope gales. Weaver, however, specializes in trying to outguess rogue winds. He had come from the National Severe Storm Laboratory in Norman, Oklahoma, where he had spent most of a decade chasing tornadoes. Twisters are still his first love and the focus of his research. But the warm, careening gales in the lee of the mountains caught his interest right away. "Downslope winds may be very localized," he says. "You often don't feel them nearly as much on the east side of town as you do right here by the foothills."

Weaver is a slim, tanned Westerner who juggles his teaching and research responsibilities with a second vocation as a volunteer photographer for the local fire department. On the wall above his desk is a montage of photographs of lightning, tornadoes, and fire destroying houses and cars. The shots are at once chilling and beautiful.

During his first year in Fort Collins, Weaver began talking to longtime residents about extreme downslope winds — which he never refers to as chinooks. Nearly everyone he consulted told him that downslope winds begin after dark. "It's a very common perception," he says, "and it's wrong. Downslope events can, and will, start any time of day or night. When people are at work, they're usually in engineered buildings, and they're busy. But you're going to pay more attention to the wind when you're home, especially if it's strong enough to shake the whole house."

Once a severe wind begins to blow, the tension it causes increases with time. "After you've lived here for a while you realize that you always get damage with a severe downslope wind; you can't avoid some destruction to fences and trees and buildings," he says. "In a tornado people suffer mostly from the sudden and dramatic nature of the event. It's really fast, and the destruction can be unbelievable. People are stunned afterward. A hurricane wears down your resistance. The wind demands every bit of your attention. You have time to worry about your safety, but it doesn't go on long enough for you to develop the kind of acceptance syndrome that you get in a war. It's somewhat the same with downslope wind events."

Might it be possible to predict such winds? In 1984 Weaver and a colleague, John Brown of the federal Forecast Systems Laboratory in Boulder, set out to analyze the weather conditions that preceded 110 episodes of severe winds. "We looked at previous research and a massive amount of data," Weaver says. "We found two scenarios for severe downslope winds. In one, a lee trough forms just east of here." A lee trough is a large region of low pressure created by strong west winds passing over a north–south range of mountains. The deepening depression acts as a sinkhole and pulls in air. Winds that have been tossed upward by the mountain peaks accelerate in speed as they descend. Weaver describes this kind of flow pattern as "an almost self-feeding beast."

Lee troughs produce downslope winds of moderate strength. The strongest winds seem to form when a high-pressure system southwest of Colorado pushes air over the Rockies while a low-pressure system to the northeast — over, say, Nebraska or South Dakota — draws it eastward. "Air under high pressure basically rushes to fill the low," Weaver says.

But even in ideal conditions for downslope winds, not all foothills communities will be assaulted in equal measure. "There'd be times when Boulder would get severe events and we wouldn't," Weaver says. "It took us a while to realize that the direction of the wind can make a tremendous difference." The Medicine Bow Mountains shelter Fort Collins from winds that blow from due west. "But if the winds come in from the northwest, they miss the Medicine Bows completely, and we get clobbered."

Weaver and Brown designed a downslope wind prediction model for the National Weather Service that was first put into use in 1989. Over the years they have continued to refine it by adding such factors as precipitation and cloud movement. "We started with a critical success index of 23 percent," Weaver says. "Now we've gotten it up over 62 percent. We're still missing 20 percent of the events, and we have false alarms 20 percent of the time, but we're getting there." Such forecasts are important for scheduling agricultural burns and other outdoor activities, as well as for warning front-range residents to look out for falling air.

WHAT OF THE deadly funnel-shaped winds that strike hundreds of times yearly in the heartland of North America? Tornadoes exert a powerful hold on the imagination of Americans. They may draw us into visions of darkness and fury or transport us to the colorful world of Oz.

When I was about ten and first interested in the world beyond my own town, a tornado tore through a community somewhere in the Midwest and caused extensive damage. I can still see the pictures on the front page of the evening paper: aerial shots of shredded buildings, grim street scenes with residents sobbing over their lost homes. I wondered who would ever choose to live in a place where such things could happen. Never, I thought to myself with a shiver. I will never live in the country of tornadoes.

I have since learned that tornadoes are beautiful, fluid oddities. The chances of being caught in the path of one, even in the most tornado-plagued states, are quite small. Most Midwesterners, understanding this, wait until the last possible moment to seek shelter, even when weather forecasters warn that touchdown by a twister is imminent. Since the 1970s, improvements in forecasting techniques and equipment, particularly a new generation of Doppler radar systems, have enabled meteorologists to give much earlier warning before a tornado drops from the clouds. (However, scientists still do not understand why some giant rotating thunderstorms spawn tornadoes while others do not. Nor do they know precisely how a tornado's circulation begins or how fast it may blow.) "In the seventies," John Weaver says, "the average tornado warning lead time was minus thirty seconds. The warning would reach the public thirty seconds after the first tornado touched down. Now it's plus nine to fourteen *minutes*, and it's going to increase.

"The population density in Tornado Alley (from Nebraska, Kansas, Oklahoma, and Texas east to the Appalachian Mountains) has gone up two and a half times, and the death rate from tornadoes has gone down. That represents real success. In fact, one of our problems is trying to figure out how far ahead we should warn people. If you put out a warning forty-five minutes in advance, they may just go to a shelter, get bored, and come out."

Although Weaver's words are reassuring, I continue to wonder about the psychological toll of living in tornado country. What of people who have been trapped in the path of a killer funnel? How have their lives and their view of the world been changed?

THE WEATHER in central Iowa on October 14, 1966, was unusual by any standards. With temperatures in the eighties and a dripping wetness to the air, it seemed more like midsummer than midautumn. Five hundred miles to the west, an early winter storm had buried the Rockies beneath several feet of snow. Now the system was rolling eastward with long tongues of cold air. As it stabbed into the heat of the eastern plains, the front created perfect breeding conditions for tornadoes.

Of the several twisters that formed, all but one spun rather harmlessly through cornfields and isolated patches of woods. But in midafternoon, just southwest of the small town of Belmond, a major vortex dropped to earth, and not with a narrow, jumping tail. The whole wide body of the funnel alighted and traveled through the center of town.

It was the start of Homecoming Weekend in Belmond. That evening the high school football team, undefeated for thirty consecutive games, was scheduled to play nearby Lake Mills High School. At 1:30 P.M. the town's three schools let out so students of all ages could attend the annual parade on Main Street. By 2:15 a pep rally was in full swing in the heart of the three-block shopping district. A half-hour later the crowd dispersed, in part because of the clouds that loomed on the western horizon.

Maxine Dougherty had just arrived home from her husband's jewelry store on Main Street when she saw a black mass approaching from the southwest. She stared at it through her front window. Tornadoes didn't usually form in autumn, but within the cloud she could see several skinny tails waving and dancing. She ran to the phone and called her husband. "Get everyone in the basement," she cried. "A tornado's coming, and it's headed right for you."

In a Main Street beauty salon, where Sandy Anderson was having her hair done, an odd, eerie light shone through the windows. "One of the other customers said, 'It feels weird out, like something's going to happen,'" Anderson recalls. "We all started teasing her, saying 'Ooh, it's Halloween time. There are ghosts and goblins in the air!' But then the wind gusted real hard and the beautician said, 'Sandy, I think we'd better go to the basement.'" The women walked quickly through the hairdresser's house, which adjoined the shop. As they went down the staircase, the back door exploded into the house.

At the jewelry shop, Russ Dougherty pushed three women customers down the basement steps and told them to hold onto the handrail. "They were just screaming and crying," he says. "All I could think about was getting my diamonds out of the front window." He filled his pockets with jewelry and ran back to stash them in the basement. As he returned to the window for a second load he saw the old brick hotel across the street crumble away. "My building was just shaking like crazy, and all I could do was stand there like a dummy," he says. "Debris was flying everywhere. I said, Oh my God. And then my windows blew out."

A few blocks to the south, Mayor Bob Misner, a veterinarian, and his daughter, Laurie, noticed the cloud a few seconds before the neighbor's garage collapsed. The mayor jumped in the air as a wave of debris crashed through the front of his veterinary clinic and covered the floor. "That was just the first big gust," he says. He hustled his wife and daughter and two visitors into the bathroom.

"There was such a terrific shaking and roaring," Misner's wife, Eleanor, recalls, "but then a lull came and lasted for about twenty seconds." The passing of the eye. "I laughed, from nervousness I guess, and my daughter got mad at me. Then the racket started up again."

Younger students had been taken back to their schools after the pep rally. On the first floor of the junior high school, Russ and Maxine Dougherty's twelve-year-old son, Steve, saw a dark mass roll across the town. By then everyone knew a storm was coming. A janitor was supposed to lead the class to the school basement, but he never appeared. As Steve watched, trees toppled over, their roots spewing dirt. Buildings flew apart. A car flipped upside down. The cloud advanced across the playground, sweeping aside swing sets and jungle gyms. "It sounded like a semi coming down the hall," Steve Dougherty says. "You couldn't hear yourself think." He crouched by the windows, head down and trembling, as the storm swept over the school.

It was over in less than five minutes, though it seemed much longer.

Emerging shakily from the hairdresser's basement, Sandy Anderson felt as if she had entered an utterly unfamiliar world. "The trees were shredded, the buildings were gone. There was debris hanging everywhere. And there was this morbid quiet. We all just kind of stood there, looking down Main Street. It was like being in a war.

"Finally I heard a voice coming from the building next door. I'll never forget it. It was a woman, and she said, 'I know I'm not the only person alive. Somebody say something.'"

ONE OCTOBER AFTERNOON, exactly thirty years after the Belmond tornado, I drive north from Des Moines through fields of dried corn and scattered groves of maples turning from green to gold. A gilded land. It is a surprisingly hot day, unseasonable and muggy, with tomorrow's forecast calling for fierce cold and wind. The parallel is not lost on me. Every few minutes I glance toward the western horizon, which remains reassuringly clear.

I am going to Belmond to hear residents recount their experiences in the tornado, as I am told they love to do. Six people were killed that day. Seventy-five businesses and 109 homes were demolished.

Another 160 houses suffered major damage. But my purpose is not simply to gather eyewitness accounts. I want to know how living through the storm changed the residents' lives.

I learned of the Belmond tornado through Mary Swander's wonderful book *Out of This World.* In her chapter about the Iowa spring, Swander describes the pervasive and wearing quality of the Great Plains wind, which trundles in from the west like a never-ending train. It begins, she writes, early in the day, "quietly at first. . . . Then as the morning comes into full adolescence, its dance card fills, the wind its steady partner. . . . I hang on to the flagpole and fence posts to make my way from house to barn, little bits of hay stubble flying up into my face." I know how Swander feels. Living in so breezy a landscape is difficult from a physical standpoint, yet so bracing as to be worthy of celebration in words. But as I began reading Swander's description of tornadoes, a sudden, tingling fear arrested my breath. She writes of a brief teaching assignment she held in Belmond ten years after the tornado. "This town had lost not only most of its physical possessions but its architectural history and the memories, the landmarks — both real and psychological — that accompany a sense of place. The trauma of the event still lingered. Children who hadn't yet been born when the tornado hit had nightmares about the disaster. They wrote about them in their poems."

Belmond is a one-horse, one-light town founded in 1856 and bisected by U.S. Route 69. At 2,500, the population is the same as it was before the tornado. A welcoming sign lists a number of community service organizations and nine churches. On Main Street, where stately nineteenth-century brick buildings once stood, the well-kept one- and two-story businesses are threaded together by a flat turquoise roof over the sidewalk, supported by pinkish metal stanchions. In the gathering dusk globe-shaped streetlights burn a soft ivory. I wonder how the townspeople feel about the shopping district's flashy modern arcade, designed in the late sixties. It is unusual and attractive, though I would have preferred the original, historic buildings. Do residents regret the loss of the town as it was before?

At seventy-three Russ Dougherty has curly, graying hair and a solidness to his manner — his handshake, his gaze, his voice. When we meet for breakfast the morning after my arrival in town, he insists on giving me a guided tour of the tornado's path. He begins by pointing out the town's major employers: a seed company, a thriving print shop, and southeast of downtown a plant run by the Eaton Corporation that manufactures machine valves. The town gem. Depending on the times, the plant employs between 600 and 700 people. But Dougherty adds, "The saying around here is still, 'As goes farming, so goes Main Street.'" A year of poor crops means trouble for everyone.

Dougherty was one of the main forces behind the development of the arcade on Main Street. "Our theme was 'Rebuilt and ready,'" he says, "and we meant it. When the arcade opened, the interstate hadn't been built yet. Route 69 was still the main drag from Minneapolis to Des Moines. People would pull up to that stoplight, look down Main Street, and really be impressed, especially at Christmastime when we decorated trees and put 'em on top of the roof. We got all sorts of positive comments and letters."

"Do you ever miss the old buildings?" I ask.

"Not really," he replies. "They were in pretty bad shape. I think the tornado gave us a stronger, better community. People had to pull together, and it made us appreciate each other."

We drive slowly through the northeast quadrant of town, the residential area hit hardest by the storm. After the storm the town was virtually treeless, but now lawns are covered with the yellowing leaves of maples thirty and forty feet tall. I notice a number of grand two- and three-story houses with high front steps, sloping roofs, and pretty porches, built early in the century. Set between them are simple one-story ranch houses dating from 1967 and 1968. A few vacant lots show where houses were never rebuilt. "It was the darnedest thing, seeing which buildings got destroyed and which ones didn't," Dougherty says. "See that?" He points to a shabby wood-frame house with cracked asbestos shingles. "It was hardly touched. But right across the street was a really nice old place, and it was a total loss."

I trace the path of the storm through architecture. The zone of destruction measures seven-tenths of a mile in width. Along First and Second avenues, every house is new. "This area just got plastered," Dougherty says.

"The really wonderful thing was the outpouring of support from other communities," he tells me, more than once. "It wasn't but a couple of hours before we were flooded with people coming in to help. Food, donations, you name it. And the people from here all pulled together, too. It was unbelievable. It really gave us a sense of cooperation and sharing. Even the churches got together, and before — believe me — the churches all went their separate ways."

OVER THE next two days I have no difficulty finding people who are eager to talk about the storm, about where they were at 2:55 that afternoon. About what they saw, what they felt.

A gas station collapsed and blew away, but a car remained evenly balanced on a lift in the station bay. A tractor-trailer flipped up on one end and danced out of town, waving like a flag. The glass in the windows of Main Street shops bowed in and out, as if made of rubber.

A local mailman, walking his route, was entombed by a metal storage bin that wrapped around him. Discovered when someone noticed his shoes sticking out from beneath the twisted metal, he survived.

People caught outside were covered with mud and dirt that took them days to scrub off. Leaves hitting the bare legs of schoolgirls felt like flying glass. Teenagers held hands to keep from blowing away.

In the near-dark following the storm, firemen searched door to door for victims. Downed electrical lines lay everywhere, but because the storm had destroyed a power substation, none of them were hot. Nurses gave people tetanus shots on the street.

A cold rain beat down that evening, and overnight it snowed. The air smelled of dust and old insulation. "You can't believe the mess," people say over and over. "It was like a war zone. It was beyond your wildest dreams."

I am told also of the crazy places that possessions landed, and of funny things people said. The bankbooks belonging to Sandy Anderson's two children turned up in a neighbor's bathtub. A woman whose house was badly damaged found an artificial apple and orange beneath the spread of her still neatly made bed. A tire from the gas station somehow made its way into the basement of a nearby house, though it was too big to fit through the window. "My place wouldn't have been in such bad shape," one woman jokes, "if my neighbors had kept their stuff at home."

A mother sifted through the wreckage of her house in search of the birthday cake she had just baked for her daughter. A young man, finding nothing left of a friend's house, shouted desperately into the basement, "Luke, Luke!" Luke's father popped up like a jack-in-the-box and yelled, "Luke ain't home!" That story always sends its tellers into peals of laughter.

Everywhere I go I ask the same questions. Do tornado warnings scare you now? Most people chuckle and say that when a siren sounds, they go outside to watch the sky. What do you remember about the cleanup? I ask. "That's all a blur," people typically answer. "It's been so long. But the day of the storm — *that* I remember, clear as if it were yesterday." Residents echo Dougherty's sentiments by the dozens. The tornado made them pull together. The town became stronger because of it. "Just look at what we got out of it," says Don Cleveland, a retired school superintendent. "Look at the arcade. People say, Boy, if you could get everyone out of danger, a tornado might be the best thing that could happen to a town."

This unrelenting enthusiasm makes me suspicious. When I ask to speak with someone whose house was lost, I am directed to Sandy Anderson, who now works at the town library. In 1966 Anderson was a young widow with two small children. Her husband had died the previous year from a kidney disease. The tornado damaged her house beyond repair.

I find her to be a pleasant, well-spoken woman in her fifties who exudes good will and talks about the storm easily. "It was a difficult time, yes," she says. "I was sad to lose my pictures. But I lost *things*.

My kids and my parents were okay. There wasn't much reason to grieve. Not at all like when I lost my husband."

What about the town's children? I ask. Weren't they scared? Wasn't anyone affected badly by the destruction, the turmoil?

"The children I knew were pretty well protected from everything by parents and grandparents," Anderson says. "And the teachers did a superb job of helping them feel safe." She pauses. "Maybe if you talk to someone who was injured. I know a woman who was caught with her two small sons when a wall caved in on her."

A few minutes later I cross Main Street and enter the P&G Supermarket, which houses a delicatessen where a good portion of the town eats lunch. In the deli I ask for Nancy Schimp. She is a stout, curly-haired woman seated at a small table. As I introduce myself, she shakes her head.

"Lots of people know more about the storm than I do," she says firmly. "I don't remember anything. We were taken to a hospital out of town, anyway, so we missed all the cleanup."

"But that's why I want to talk to you. Because you were hurt."

"Lots of people can tell you more than I can." She glances around uncomfortably. Our conversation has attracted the attention of people at surrounding tables. I stand awkwardly beside her, wondering what to do. "I don't even like to talk about it," she says, her voice growing suddenly soft, her hand coming up to her chest. "It still terrifies me."

When Schimp saw the black mass of the approaching tornado, she grabbed her three- and four-year-old sons and retreated to the southwest corner of her basement, the corner facing the storm. "That's where they always tell you to go," she says sarcastically. And usually that is the safest spot, because houses tend to blow away with the storm. But as the winds subsided, Schimp found herself and the boys buried in concrete rubble. The basement wall had collapsed. "The oldest one got out right away and started running around upstairs, barefoot, with all the broken glass and downed wires," she says. "The youngest one was really buried. I thought he was gone.

"My leg was cut, and I broke my collarbone heaving a piece of concrete off my youngest. He was in the hospital five days. Had a bad concussion." She is talking louder now. I have hardly said another word, but she seems bent on telling me her story. For years, she says, the sight of storm clouds sent her into a panic. She never again felt comfortable letting her children out in the yard to play, "even on a sunny day. I just had this terrible fear that it might happen again. It lasted until they graduated from high school."

I do not press her to tell me more.

That evening I call Jeff and relate Schimp's story. I say I find her early reticence surprising. "Everyone else has been so open. It's almost like it does them good to talk about what they went through."

"She just got more shaken up by it," Jeff replies. "Think about it. Think about how excited people are when a hurricane gets close to the Outer Banks. Not me. I just get quiet." He's right. When Jeff was a teenager, Hurricane Camille destroyed his hometown on the Mississippi coast. "I know what kind of power hurricanes have. They don't excite me. They scare me."

Psychologists say windstorms may indeed leave lasting scars, although it is easier for people to recover from such natural disasters than from catastrophes like plane crashes or terrorist bombings, which strike them as utterly senseless. Residents of cohesive, family-oriented communities seem to fare better psychologically after windstorms than people living far from their families or alone in big cities. Those who are mentally unstable by nature have the most difficult time putting their lives back together. But in a study of victims of a 1989 tornado in Huntsville, Alabama, therapists found that many well-adjusted people also suffered prolonged mental distress. For years afterward they would become extremely nervous in high wind.

The day after my conversation with Nancy Schimp I find a dozen people who admit to being continually spooked by dark clouds and tornado warnings. Mary Jo Swenson, a customer in the P&G, tells me, "I go down in the basement at the first siren, with as many of my precious things as I can carry. My daughter is absolutely terrified of storms."

"I still have nightmares a couple of times a year," says Randy Covington. "There are four or five twisters coming, and no place to hide, and I've got to weave my way through 'em."

"Last summer I went to see the movie *Twister* and cried all the way through it," says Charlotte Gabrielson. "I guess I thought I was completely over the storm and that it would never bother me again. Now I know I'll never be completely over it."

Kathy Goeman found herself sobbing over a newspaper story commemorating the tornado. "It took me right back to that day," she says. Among her friends she is well known for her fear of wind. "All wind irritates me," she says, "but storm wind scares me." Her pulse speeds and she begins to pace the house. A lead ball forms in her stomach. When I ask what emotions she feels, she thinks for a moment. "Panic and fear," she replies. "I see everything about that day — how beautiful it was, how dark the sky got. And then, afterward, people walking around bleeding, saying, 'Have you seen my children? Have you seen so-and-so?' "

Goeman was twelve when the tornado hit. She took refuge in Belmond's tiny hospital. When the electricity went out, a patient on a respirator died, "and all the nurses started crying. Then they started bringing in victims from the storm, and it was awful. There were Red Cross volunteers running around in a panic. They had been trained for disasters, but they'd never seen anything like this.

"On the way back to our house we passed a farm, and there were men shooting pigs to put them out of their misery." The animals had pieces of wood jammed into their sides. Near home, she watched searchers root through the rubble of a house and pull out the corpse of an elderly woman.

"I've ruined my children," she says. "They're scared of storms too; they got it from me. I keep thinking all this will fade in time, but it hasn't."

I wonder silently if Goeman's experience was more disturbing than that of other townspeople or if she simply has a better memory.

For the rest of my visit people regale me with stories of the storm, many of them laced with humor. A mother tells me of her son "who was hellbent to get away from this podunk town" but ended up dropping out of college to help with the cleanup. He lives in Belmond still. No one, not even those who suffer from fear of high winds, expresses bitterness about the storm. Hardly anyone even expresses regret. The old buildings are missed, but they were in disrepair anyway. When a tornado destroyed part of nearby Charles City a few years later, Belmond sent busloads of volunteers to help with the cleanup.

In the 1960s few people went to counselors to work through their personal traumas and fears. Instead they told stories to each other, over and over. In Belmond they are telling them still. In 1988, when a mother and daughter advertised in the local paper for accounts of the tornado to be published in a book, they were flooded with manuscripts. "Talking about it helps with the healing," Norma Jeanne Jenison, the mother, tells me. "It's the same as grieving over a death. And the overall story is one of strengthening. We're a fighting town, and we survived."

The outcome might have been much different, Jenison acknowledges, if the tornado had touched down when Main Street was still filled with young revelers. As it was, the six people killed ranged in age from fifty-nine to eighty-two. "We were lucky. Everyone recognizes that," she says. "If we'd had to bury many little children, this town might have died instead of coming back to life."

IN TORNADO ALLEY terror drops from the clouds with a few scant minutes of warning and plows a single furrow across the land. Here on the Atlantic coast it comes in a larger, more ponderous form that leaves ample time for preparation and worry.

It is summer again on the Outer Banks. We awoke this morning to a yellow light and a low, dull layer of clouds. Outside, the beach towels on the clothesline drip with humidity. A salty wind blows off the ocean, but the day is steamy hot, and the weight of the air presses against my senses. My thoughts are muddled, my nerves on edge. Four hundred miles to the south, a circular tempest wobbles toward the East Coast. It is Bertha, the first hurricane of the season.

With Jeff at his office and our son at a sitter's, I am supposed to be hard at my writing. But it is late morning, and I have not been able to make myself form a single sentence. I go into the kitchen for the third time and turn on our weather radio. Bertha is a high category-two hurricane on the Saffir-Simpson damage-potential scale, with winds to 105. It is not supposed to hit the Outer Banks but to go into more populated Wilmington, two hundred miles to the south. All week I have had a strong intuition that Bertha will not come here, but in the last few hours my feelings have begun to change.

Arching as far east as they do, the Outer Banks appear on satellite maps as a finger beckoning to tropical storms. The hardy souls who first peopled the exposed strands of Hatteras and Ocracoke islands learned that in summer and autumn their footholds on terra firma might suddenly disappear beneath ocean. Houses were built to float away rather than break up during floods; to this day a few island residents own particular parcels of land simply because their residences happened to come to rest there after a storm. Then in the early 1960s the pattern changed. For more than two decades the Atlantic seaboard was spared as storm after storm swept into the Gulf of Mexico. The population of sleepy little coast towns quickly grew, despite the warnings of meteorologists that the pull of hurricanes to the north would someday resume.

In 1984 Hurricane Diana came ashore in North Carolina a bit south of the banks. In 1985 I moved to Hatteras Island six months before it took a direct hit from Hurricane Gloria. Hurricane Charley followed in 1986, and the devastating Hugo, which made landfall in South Carolina, in '89. Bob spun past in '91; Emily whirled across Cape Hatteras in '93. I speak here only of storms that made it far enough north to threaten our neck of the woods. The hurricane I think of as the Big One, Andrew in '92, leveled entire communities in a band across southern Florida and Louisiana. Although the death toll was small — only twenty-three killed — the people in Andrew's path will long be haunted by the memory of the raw power that swept through their lives.

We do not take hurricane warnings lightly on the Outer Banks. We study the characteristics of the storms as they approach, especially their winds. If a system has sustained winds over one hundred miles an hour, our family and most of our friends evacuate; anything less we can ride out. Or so we believe. We watch as each storm wobbles along, and we fret about what will happen when it crosses the Gulf Stream, the hot northward flow that serves as a bloodline for tropical storms. To some degree our ability to predict the path and destruction of each tempest is a measure of our wisdom. Island natives scoff at people who run from storms. Who among us newcomers is savvy enough to know when a hurricane is truly dangerous and when it is run-of-the-mill?

Bertha is one to be watched. When it was still hundreds of miles from shore, the winds near the eye reached a maximum sustained speed of 105 miles an hour. Then it lost strength, and I dismissed it. But inside the Gulf Stream it regained power and leveled its aim directly at our stretch of coast. If it does not come ashore at Wilmington, it is expected to drift north, make landfall at Cape Lookout, and move up Pamlico and Albemarle sounds. The latter course would put us in the northeast quadrant of the storm, where the hurricane's strongest winds are combined with its forward movement: a storm with 105-mile-an-hour winds moving at 15 miles an hour packs a 120-mile-an-hour punch.

I do not really believe it will hit here. Outside, patches of blue are showing through the clouds now, and my instincts tell me we are safe. Nonetheless, this is a big storm. Landfall, wherever it is, will come sometime this evening. We still have time to pack up and leave.

Indecision grips me like an illness. To stay means a long, windy night with no electricity — this is almost a certainty — and branches crashing onto the house. Our son might be terrified. But what if we drive west and Bertha comes ashore in Wilmington and then moves inland? Might we only be moving further into its path?

A HURRICANE or cyclone or typhoon (all are the same) is born as a small disturbance in an ocean with surface temperatures of more than eighty degrees, where surface winds from different directions converge and a high-pressure system in the upper atmosphere draws air skyward. These conditions are met only in the open oceans of the tropics. Warm, humid air fosters unstable conditions. Winds in the upper troposphere, blowing uniformly from one direction, enable advancing bands of thunderstorms to begin a snakelike coiling. That is what turns within a hurricane; hidden by a skirt of clouds is a series of concentric thunderstorms, each three to thirty miles wide, each spiraling inward toward the eye.

Unlike the midlatitude storm systems known as extratropical cyclones, the winds inside a hurricane are strongest at the earth's surface. Air spirals upward within the storm, reaching its fastest speed on the wall of the eye, then descends downward in a cloudless chute.

When the thunderstorms first become organized into a swirling mass, the National Hurricane Center in Miami announces the creation of a tropical depression. Once winds within it reach speeds of thirty-nine miles an hour, the system is declared a tropical storm and given a name. At seventy-four miles an hour, it becomes a hurricane. As long as it moves and breathes over tropical or subtropical ocean, it is likely to strengthen, for hurricanes feed on hot, humid air. Once it starts plowing across land, however, it loses strength as quickly as if it has run into a snowbank. As a result, inland communities are usually much safer than those on barrier islands — though during a major storm, cities two hundred miles inland may suffer torrential rains and winds of seventy-five miles an hour.

The Saffir-Simpson hurricane damage potential scale places individual hurricanes in five categories based on their wind speed, storm surge height, and barometric pressure. The only category-five storm to have come ashore in the United States since the 1930s was Camille, which assaulted the Mississippi coast in 1969 with winds of more than two hundred miles an hour. Whenever members of Jeff's family describe the damage wrought by Camille, their voices have a deeply somber tone. The family business was destroyed, as was the house of a dear aunt. To me the most haunting image Jeff's siblings describe is of a chainlink fence stuffed full of dead songbirds.

My own firsthand experience with hurricanes began with Gloria in 1985. With winds up to 130 miles an hour, Gloria was a category-three hurricane, and at that time the largest storm ever to develop in the Atlantic. Thirty-six hours before it made landfall on Cape Hatteras, a heavy layer of clouds obliterated the sky. The air felt ominously heavy and still. A leaden mantle settled onto my shoulders; I knew the storm was bound for my island, my home. I packed my most cherished possessions and fled. But Gloria lost power, came ashore at dead low tide, and traveled up the cape so quickly that the full extent of her rage was never felt.

Since then I have come to depend heavily on instinct when we must choose whether to evacuate. I base my decisions on the look and feel of the sky, alert always for the odd, foreboding energy that heralded Gloria. In 1989, when Hugo approached the Carolinas and comparisons were made to Camille, I stopped Jeff from loading our getaway car. The skies were blue, the winds light from the north. "It's not coming here," I said. "Let's stay put." My parents called and urged us to leave, but I stood firm. Hugo blew in south of Charleston, traveled far inland, and flooded roads in the communities to which many Outer Banks residents had fled — while Jeff and I enjoyed a candlelight dinner beneath the stars.

Storms are an integral part of life on the Outer Banks, a source of exhilaration and pioneerlike toughness. They bring us together; they slap us awake. As a hurricane becomes a palpable threat, men in the streets mention casually that they've got to go haul their boats out of the water. A damned nuisance, they say — but their tone of voice belies their excitement. Women, bright-eyed, stop each other in stores and schools to share what news they've heard. That, at least, is how hurricane season begins. But when a major storm bears down on the banks, our minds grow numb. We go about our lives in a mild state of shock seasoned with dread.

IT IS FOUR o'clock in the afternoon, and Bertha is not doing what forecasters said it would do. Instead of veering south toward Wilmington, the latest coordinates reveal a wobble to the north, which would bring it straight in on the Outer Banks and up the sounds. The winds stand at 105. The official word is still that Bertha will assault Wilmington, but several dissenting voices have the storm looping our way. Should I call Jeff at work and tell him I want to leave? An electric current begins to run through me, born not of fright, really, but of an acute awareness of my surroundings. I still do not believe Bertha is coming north — but who am I to try to outguess a loose cannon?

I go outside and stand in the yard, trying to regain my intuition. Patches of blue still show through the clouds. We've got it made, I think. Nevertheless, back in the house I take photograph albums and my most precious files from my second-floor study and stash them underneath the table in the dining room — just in case a tree falls on the roof.

Jeff arrives home at five-fifteen. He looks at me once and says, "What's wrong with you? We're *fine,* especially back here in the woods." My mind is flying. Should we park the cars on the street or take the chance of a tree falling on them in the yard?

When we go to pick up our son, the baby sitter says, "The old-timers are saying this one's coming up the sounds."

"Really?" I say.

"Well," Jeff answers, "we're here now, like it or not. The bridges are closed; they're not letting anyone on or off the islands." On the way home he says, "The locals on the mainland are saying the storm's going inland, over them. It's a crapshoot, Jan."

The winds are only about thirty miles an hour, still northeast. At home I finish filling the kerosene lanterns and turn on our weather radio. "All preparations should now be in place," a woman says, with an edge to her recorded voice. "In case of emergency, remember that medical care may be delayed for several days. Bertha is a dangerous hurricane with winds . . ." I turn it off.

By seven o'clock the gusts through our trees have climbed to fifty miles an hour, and the skies are turning dark. I look up to the crowns of the loblolly pines, circling like gyroscopes sixty feet overhead. Their movements are mesmerizing. I will not be able to see them much longer in the failing light. With food on the table, I sit down and gaze up through the window. "Ahem," Jeff says. "It would be nice if you ate dinner with us instead of with the sky."

After our meal Reid goes off happily to play, while I pace the house, looking for something to do. Gusts seep beneath the windows and doors. The wind is a dragon exhaling in bursts. The wind is a kettledrum being played above us with dull, rolling strikes that rise to deafening crescendos. My parents call from Delaware, and I assure them that we are fine — wishing all the while that we had left.

At nine o'clock I retreat to the back porch with the weather radio, which is filled with static and barely audible. As I turn it on, I think of women who listened in secret to news of World War II to keep from upsetting their children. This is a war I am part of, and tonight I am losing the battle of nerves. Small branches pelt the roof like shells.

Bertha has made landfall north of Wilmington. Its course now is anyone's guess, but it appears to be headed inland, away from the coastal sounds. To my utter disbelief, our power has not yet failed. At ten o'clock my parents call to relay a report from the Weather Channel that the storm has lost strength; winds are down to eighty-five. I sigh deeply and feel my jaw unclench, my shoulders relax.

Just before midnight I step out onto the porch in gusts that I gauge to be about seventy miles an hour. The pines scream. With a large flashlight I can see that they are bending like bows. I press my back against the house to keep out of the way of falling twigs and branches. Bertha's eye is athwart us on the mainland; for the Outer Banks, at least, the storm will get no worse. The wind roars and lets up, roars and lets up — and I take in a bellyful, a breath at a time.

IF IT WERE POSSIBLE to stop hurricanes from rolling up the Atlantic seaboard, if only for a single season, I doubt that even the hardiest Outer Banks resident would say no.

It is early September now, and we are worn-out. Six weeks after Bertha, we were threatened with Hurricane Edouard but again dodged the bullet — or, rather, the bullet dodged us. Edouard stayed to the east, grazed islands off the New England coast, and dissolved. Then two days ago Hurricane Fran came ashore near Wilmington with 115-mile-an-hour winds and destroyed hundreds of homes. It continued inland, downing trees and power lines in Raleigh, dumping fourteen inches of rain in the Appalachians, and flooding Washington, D.C. We were untouched but not unshaken. Twenty-two people are dead, and more than a dozen counties have been declared disaster areas. We can't get through to our friends in the mountains; we worry that their house has been swept away by a flooded creek.

Hurricane season is barely half over.

Tonight six adults, all permanent Outer Banks residents, all veterans of many storms, sit around a groaning dinner table at the house of our friends Jim and Marcia Lyons on Cape Hatteras. To a small degree we know what faces the victims of Fran as they begin digging out, for we have all been in similar straits. The weeks after a hurricane often stay hot and muggy, and mosquitoes swarm unmercifully. Raw sewage stands in ditches. Water and ice become as precious as gold.

Jeff thinks of Hurricane Camille, and I think of Gloria. Two years ago Jim and Marcia and the other couple here, Mike and Nancy Cowal, were trapped at home for days after Hurricane Emily felled hundreds of trees. The Cape Hatteras School, where Jim, Mike, and Nancy teach, flooded to four feet. "I hate to admit it," Nancy says, "but I'm getting tired of this." I know how she feels. Everyone on the North Carolina coast knows how she feels.

"I've heard people actually suggest that the government do something to slow these hurricanes down," Mike says, shaking his head. "As if anything we do could have an effect on something so powerful." (Outlandish as it sounds, in the 1960s the U.S. Weather Bureau tried seeding the eyes of two hurricanes with silver iodide in hopes of reducing the pressure gradients within the storms and weakening their winds. Although the storms began to disintegrate, within a few days they reorganized and regained their strength.)

We are silent for a few minutes. In an adjacent room our children whoop and holler over some sort of game. "At least we don't have to shovel snow in the winter," someone jokes. At least tornadoes are rare here, and muggings, murders, and rapes. Danger lurks in every landscape, we agree. Ah, but to some degree you can protect your family, your way of life, from most forms of catastrophe. How can you keep hearth and loved ones safe from the world's most powerful storms? We all know the answer. It cannot be done. The strongest of houses and the most prudent of evacuation plans will never alter the fact that we live in a natural danger zone.

The children dissolve in a tangle of wails; the dinner party breaks up. As our family loads into the car, a light southwest wind ruffles the pines. I look up to a sky filled with stars and wish for solace for the victims of Fran. For them and all those I love, I wish a lifetime of peaceful, healing winds.

NINE

Tapping the Flow

SOMETIME IN THE TWELFTH CENTURY, as Zen Buddhism was becoming increasingly influential in Japan and the country was adopting a military style of government, a samurai warrior named Minamoto-no-Tametomo was exiled with his son to the desolate island of Hachijo Jima, a hundred miles south of the Japanese mainland. Tametomo was a crafty man. It occurred to him that it might be possible to harness the wind to transport his son back to the mainland. From unknown materials, perhaps wood and silk, Tametomo constructed a kite large enough to carry the boy over the channel known as the Izu Shichito. Sent aloft on a day of moderate wind, Tametomo's son reached his homeland safely. Or so legend holds.

I have always liked the image of kites bearing people to places they could not otherwise reach, of wind-borne scraps carrying soldiers over the walls of fortresses, lifting bricks and tiles to workers on the top stories of houses. That is what Marco Polo found when he reached China in 1282: kites being used as an upper-level delivery system.

Today I have particular reason to remember Tametomo and his son. At two o'clock I am scheduled to be strapped into a hang glider with a master pilot, towed to 2,000 feet above the coastal plains, and cut loose. Any turns I make, any swoops over the countryside, will be up to me and a master pilot.

I have long been curious about hang gliding, in part because planes and gliders are an important part of Outer Banks history. Wilbur and Orville Wright made their first flight here in December 1903, choosing these islands because of the constant wind. Francis Rogallo, the man who invented the hang glider, experimented with different designs on the northern banks. I envision myself hovering on high, a human bound to a kite, lighter than air, feeling the shape of the wind. I imagine catching a thermal and letting it bear me skyward.

I arrive at the Currituck County Airport to find the pilots way behind schedule. Three people must make flights before me. Chad, a pleasant young man with a blond ponytail, greets me, shakes my hand vigorously, and gives me a stack of forms to fill out. At the top of the first sheet is a quote from Wilbur Wright, circa 1900: "For some time now I have been afflicted with the belief that flight is possible to man."

Fifty yards away, Chad positions himself in a tandem-tow glider on a grassy landing strip. He hangs beneath the kite, parallel to the ground, in a sleeping bag–like flight suit. Cables from the back of the suit connect him to the craft's lightweight metal frame. By shifting his body, Chad will steer the glider left and right, up and down. A student pilot hangs in another bag about six inches above him. The kite consists of a piece of Dacron fitted over a lightweight metal frame to form an airfoil. Thirty feet wide, the glider sits on three legs like an unwieldy stool. Small wheels jut from each leg. As I watch, an ultralight plane revs its engine and begins rolling down the runway, pulling the glider behind it. Both are airborne in seconds.

Overhead, cumulus clouds mass and shift on their way out to sea on a westerly breeze. The buzzing ultralight tows the glider in a large rectangle once around the airport, then twice, climbing to 2,000 feet. The two craft look disconcertingly fragile, like models made of toothpicks and tissue paper. I see the plane tip its wings in a signal and fly off. Moving freely now, the hang glider pivots toward us and away, then toward us again, as if mimicking the five turkey vultures with which it shares the sky.

A dark-haired, athletic-looking man on a picnic bench aims a camera with a long telephoto lens at the glider but doesn't take a shot. "Bad light," he says under his breath.

"Friend of yours?" I ask, nodding toward the sky.

"Yeah." He smiles, still looking up, then turns to me. "We've been waiting to do this for three days. The weather's been lousy."

"You been up before?"

"A few times." He grins. "I've been hang gliding since 1974." He pauses reflectively, as if considering whether to tell me more. "At one point I did some counting and figured out that I was probably one of the first fifty people to start gliding east of the Mississippi."

I introduce myself and tell him of my interest in wind.

"Rick Sherman, from Herndon, Virginia," he says, with a firm handshake. "I haven't been up for about five years; I guess I've just been busy with other sports."

We watch as the glider wheels around, a graceful boomerang. The craft is white with a yellow stripe; against the sky it appears as the palest blue. Sherman snaps a few pictures.

"How much wind can you fly in?" I ask.

"Oh, a lot, if it's not gusty," he says. "More than you'd think. The best flight I ever had was at a place called Brady's Branch, overlooking the Allegheny River up near Pittsburgh. Middle of winter. It was twenty degrees, and the wind was twenty miles an hour but smooth. It took three people to hold the glider down so it wouldn't take off by itself. I stepped off the platform and went straight up seven hundred or eight hundred feet. It was like a giant had grabbed me by the scruff of the neck."

When a strong wind hits a cliff, it bends straight up. Hang gliders use this rising column, called a ridge band, to perform the sport's classic maneuver: hanging in one place. Gauging the width of the ridge band accurately is essential, Sherman tells me. "If you get out of it, you start dropping fast. I looked down and saw a red-tailed hawk playing about one hundred feet below me, and then I saw him turn. I followed him. I knew as long as I stayed above him I'd be in the ridge band. We played tag for about twenty minutes." Down in the Allegheny, chunks of ice floated past, a beautiful, eerie blue. "It was incredible."

"One time, on another flight, I got distracted and got out of the ridge band. I lost a lot of height, and I knew there was no way I was going to make it to the landing site. I was over a forest. All you can do in a situation like that is to look down at the trees and pick out a nice fluffy one." After landing he had to extract himself from a tangle of branches and climb down sixty feet, maneuvering his glider carefully to keep from damaging it.

The glider approaches the runway, and Sherman's friend, piloting under Chad's guidance, tips its nose up slightly to brake. As he rolls to a stop, Sherman says, "Not bad." A few minutes later Sherman himself is strapped into the glider and pulled aloft by the ultralight, hurtling face first through the air like a human javelin. The glider floats behind the plane, lifting gracefully, its skirts spread to the wind. When Sherman reaches 2,000 feet he drops the tow cable and begins to circle, weaving among vultures, a small, tipsy chip of sky.

IT IS AN odd sensation to hang parallel to the ground beneath a high-tech kite, especially when a few inches below you hangs a person you don't know. It is odder still to be flying three feet above the ground at thirty miles an hour (didn't I once dream something like this?), your whole body swinging backward like a pendulum.

On my head is a white helmet, with a chin guard that juts downward, toward the back of Chad's head. Chad turns his face to the side — back toward me — so we can talk, a gesture that in any other circumstance would seem intimate. He has very tan skin and nice eyes. "There are some straps on the side of your suit. Feel them?" I search along the sides of my bag and find the straps about hip level. "Keep your hands in there until we get all the way up. Then when I'm ready for you to fly, hold on up here." He taps the two rodlike supports that come down on either side of our heads, forming an inverted V. They are joined at the base by another rod — the third side of an aluminum triangle — with which Chad controls the glider.

In front of us the ultralight buzzes obnoxiously, a mechanical mosquito. The wind roars in our faces. "We'll be going about thirty miles an hour until we get all the way up," Chad says, nearly shouting to be heard, "so there'll be a lot of wind. Then the pilot will tip his wings, and I'll unhook the cable. We'll drop a little and slow down to about fifteen." We are already dropping and rising in waves, hitting little bumps of turbulence that I find disconcerting. On the ground I heard someone say something about the air being bumpy today, which can be both good and bad. The flight is scheduled to last about fifteen minutes. But if we can find some thermals — some lift, as the pilots say — we'll stay up longer.

At about seven hundred feet the farm country of Currituck County takes on the slightly unreal appearance familiar from airplane flights: toy houses, toy trees. The water in ponds and ditches glints like mercury. To the east stretches Currituck Sound. Beyond it is a thin line of land, the northernmost reaches of the Outer Banks.

Higher, and the air becomes hazy. It is not the light, white-blue haze I had expected but a dark blue-gray, as if a storm were brewing. The clouds above us look darker, too, and the air is refreshingly damp and cool. A clear day might have made for a better view, but I like the feeling that I am part of the atmospheric soup.

Two thousand feet. Facing east, we unhook from the plane and immediately drop five or six feet. The plane buzzes off, leaving us in the quiet of singing air. The trees below us are a deep black-green, the surface of the sound is navy blue. "Here we go," Chad says, reeling in the tow rope. "Put your hands up on the bars. I'm going to let you do the flying."

At his request I push myself forward to feel the glider dip, then shift back to bring up its front. Butterflies dance lightly in my belly. By pulling my weight to the left I send us spiraling left. To come out of the circle I pull back to the right. Chad indicates a large number 4 on the paved runway to our right. "Try turning and centering over the 4," he says. I pull hard right, then shift left. "Good," he says enthusiastically. I'm pleased with myself. The action of the glider is similar to that of a sailboat rudder. I know instinctively when to turn and when to pull back.

"Now let's head toward that 22 on the other end of the runway," Chad says. Again I turn without trouble. Below us I see three soaring turkey vultures, very small, much smaller than they appear from the ground. The sight of them reminds me of where I am, and I feel shaky.

"Keep an eye on those birds," Chad says. "If you see them circling, let me know. It means they're catching thermals."

We fly from one end of the runway to the other and back, circling until my head feels light. Our movements have a slightly off-balance quality, something like the feel of an amusement park ride, otherworldly, unreal. Many people fall in love with hang gliding on their first flight, but I do not expect to count myself among them. Again I have the feeling that I have done this in a dream.

I remember Tametomo's son and wonder if he was scared, alone among the clouds. The glider dips unexpectedly, and I decide he must have been terrified. I'm not scared, but I'm not completely comfortable, either. I try to relax my tightly arched back.

Chad continues to give me directions on how and when to turn, but I am getting distracted by scenery. I don't want to concentrate on steering just now; I want to look around. It occurs to me suddenly that we are twirling and swooping much more than I expected. What shape is the wind right now? Chad probably knows, but I haven't a clue.

"Wait a second," I say. "What happens if we just try to hang in one spot for a couple of minutes?"

"What do you mean?"

"I mean just fly straight and try to hover. Just hang."

"Okay, let's see." We keep our weight carefully centered, pointing the kite southwest, into the wind. The glider rides up as if on a wave, then falls abruptly into a trough. Chad pulls his weight quickly back. "That was almost a stall," he says.

"A stall?"

"Yeah. The only way you can just hang is if you find some sort of lift."

We are losing altitude; the earth looks more and more like something tangible. "Let's head back toward the 4 again to try to find a thermal," Chad says. What would that feel like, I wonder. A bubbling spring pushing up from the ground? I am suddenly numb with fatigue. When Chad asks me to turn left, I respond sluggishly.

We swoop over an airplane hangar and line ourselves up with the grassy runway. "Here's where I take over," Chad says. "The ground still looks a long way away, but it's going to come up real fast." I am not sure whether to be sad or glad. We descend steeply, meet the runway with a bump, and roll to a stop, our bodies again swinging backward with the force of the breeze.

JUDGING FROM ancient writings and clues unearthed by archaeologists, humankind has long taken clever advantage of wind for worship, pleasure, and work. Wind-driven prayer wheels designed to offer up chants of praise appeared in Tibet around A.D. 755. According to biblical passages, King David decorated his windowsills with Aeolian harps that were played by gentle breezes. In 1650 Athanasius Kircher, a Jesuit monk from somewhere in Europe, fitted a wooden box with gut strings and placed it so the wind could vibrate them. The device worked so well that it became a popular garden accoutrement throughout Europe. Americans, too, built Aeolian harps; both Thoreau and Emerson wrote of them, and a character in Robert Louis Stevenson's *The Beach of Falsea* employs one to bewitch and frighten beachcombers. But the notes produced by Aeolian harps are not loud, and the devices faded into obsolescence as cities and towns began to thrum with noise from factories, streetcars, and automobiles. In the late 1980s, when a British musician named Roger Winfield decided to build an orchestra of Aeolian harps six feet tall and taller, he found he had to fit them with microphones and tiny electronic amplifiers so they could be heard over the urban background noise. Winfield eventually produced an album of mostly eerie compositions, entitled *Windsongs* and recorded in England and Spain.

The power of wind was most commonly tapped in ancient times for sailing, but some cultures also employed it for industrial purposes. The Romans built cairns for smelting iron on west-facing hillsides so prevailing breezes could fan the fires; the Incas used a similar technique for smelting silver. In the late 1980s, archaeologists working in southern Sri Lanka discovered in Samanalawewa a cluster of sophisticated furnaces, built between the seventh and the eleventh century, that employed monsoon winds to keep the internal fires hot enough to smelt iron. During that period India and Sri Lanka produced the most prized irons and steels in the world. The researchers found forty-one stone and clay furnaces built on cliffs, each about two feet high and six feet long, with a front wall that faced into prevailing winds. Tapered clay pipes pierced the wall, creating a strong inward draft. When the archaeologists built two replica furnaces, they were surprised to find that the size and placement of the pipes produced a smooth intake of air, even in gusty winds. The local people had found a carefully controlled way to smelt iron; assuming that the region contained many other such furnaces, they may have produced as much as ten tons of high-carbon steel — a tremendous output — each year.

The most common way of harnessing the wind to perform work, of course, has been through windmills. Historians are not sure where the windmill originated, although it was known to be widely used in Persia by the tenth century to grind corn. The design of the first machines consisted of a horizontal wheel of sails, like a carousel, which turned a spindle that rotated a grinding stone. Although they could easily catch the wind from any direction, the horizontal mills were only a third to a fifth as powerful as those with vertical sails.

Historians believe the windmill may have been introduced to Europeans in the eleventh century, during the Crusades. Its spread across the feudal landscape was explosive. Medieval manuscripts contain illustrations of post windmills with four vertical sails — the classic windmill design — mounted on buildings that could be turned to catch the wind from any direction. In 1191 a British knight of Bury St. Edmunds refused to pay tithes to the Catholic church on the proceeds from a mill he had built, perhaps the first in England. His challenge to the church enraged the local Catholic authority, the abbot of Ramsey, who owned a number of animal-powered mills, apparently the only other mills in the region. The knight argued that everyone should have free access to the wind. The abbot replied, "I thank you as I should thank you if you had cut off both my feet. By God's face, I shall never eat bread till that building be thrown down. You are an old man, and you should know that neither the King nor his Justiciar can change or set up anything within the liberties of this town without the assent of the Abbot and the Convent." The abbot threatened to have his servants tear down the mill, but the frightened knight hastily ordered his own servants to demolish the structure.

But the use of windmills could not be stopped. During the thirteenth century they were found in increasing numbers throughout eastern England and northern Europe, and historians believe they reached China around 1230. At some point during this period the Dutch instituted a "wind brief," a tax paid to the lord or king over whose fields the wind blew before reaching a mill. As windmills became common, one English lord complained that they offered "quick-witted peasants an opportunity to evade manorial regulation, act independently, and become quite prosperous." This was likely an exaggeration, since few peasants possessed the means to build mills of their own. And many lords imposed a mill soke, or control of milling rights, for all corn raised on their land. In some regions, however, peasants could take their harvest to private mills that operated outside the manorial system.

Windmills greatly improved living standards in the feudal world simply by freeing women from the task of grinding grain by hand. The combination of animal power, water power, and wind so changed the daily lives of peasants that the historian Lynn White writes, "The chief glory of the Middle Ages was not its cathedrals or its epics or its scholasticism; it was the building for the first time in history of a complex civilization which rested not on the backs of sweating slaves or coolies but primarily on nonhuman power."

The windmill reached its apex in the Netherlands, where between

1500 and 1650 water-draining mills increased the amount of tillable acreage by 40 percent. Sails on the mills provided the power for wooden wheels, designed much like traditional waterwheels, that removed water from fields a scoopful at a time. Around 1600 the Dutch began building industrial windmills capable of sawing wood, hulling barley and corn, and stretching fibers to make hemp. Historians estimate that during the eighteenth and nineteenth centuries 100,000 windmills may have dotted the fields of Europe. But operation of the machines was dependent on the weather, and they could not compete with the steam engine. In the mid-nineteenth century, as it became possible to mill grain and saw wood with steam and to transport flour and other products from region to region by train, windmills became little more than objects of nostalgia — except in the American West. There, after 1850, settlers used a disk-shaped windmill with many blades and a taillike vane to pump underground water to the surface. Windmills enabled ranchers to water their cattle inexpensively and so played a major role in the settling of North America. In places they are still used.

WIND CANNOT be easily ignored in a world where so much commerce and travel is by air. Modern aviators must gauge the mood of the wind on every leg of every trip. Pilots know, perhaps more intimately than anyone else, how the movement of air resembles the movement of water. They seek different altitudes to find its most advantageous flow. Like rafters on a river, they fear the rapids, the places where air is churned up by the roughness of the landscape and the tug of different pressures. A range of mountain peaks, a stand of skyscrapers, a change in the vegetation that covers the earth, a massing thunderhead: all can roil the layers of air closest to the earth's surface and make takeoff and landing exceedingly tricky. At times the turbulence extends upward many thousands of feet, much higher than pilots suspect. Unusually strong winds, too, can making flying dangerous, especially for the smallest planes.

In 1940 the French writer Antoine de Saint-Exupéry published the book *Wind, Sand and Stars,* which described his experiences during eight years as a pilot flying prop planes. One day, on a flight along the coast of Argentina, Saint-Exupéry encountered a wind unlike any he had ever felt. "Here the crust of the earth is as dented as an old boiler," he writes. "The high-pressure regions over the Pacific send the winds past a gap in the Andes into a corridor fifty miles wide, through which they rush to the Atlantic in a strangled and accelerated buffeting that scrapes the surface of everything in their path." Accustomed to flying along the Atlantic coast, he thought he knew what to expect: a "grey-blue tint to the air" that revealed the zone of the highest winds and "an hour of stiff fighting and stumbling again and again into ditches of air." On this particular day, though, he was alarmed to see that the sky was pure blue — too blue. Before he knew what was happening, he was deep in the dragon's mouth. His plane began to quiver. Looking down at the crumpled earth, he realized he had stopped making headway. The ground fell away as the plane turned over. Mountain peaks loomed in front of him, then disappeared. "Vertical, oblique, horizontal, all of plane geometry was awhirl," he writes. All at once "I understood the cause of certain accidents in the mountains when no fog was present to explain them." He realized he was fighting not only the wind but the mountains. He had slammed against the flow of air over the Andes just as if he had hit a rock wall.

> There had been granted to me one second of respite. Two seconds. Something was collecting itself into a knot, coiling itself up, growing taut. I sat amazed. I opened astonished eyes. My whole plane seemed to be shivering, spreading outward, swelling up. Horizontal and stationary it was, yet lifted before I knew it fifteen hundred feet straight into the air in a kind of apotheosis. I who for forty minutes had not been able to climb higher than two hundred feet off the ground was suddenly able to look down on the enemy. . . . And at that very moment, half a mile from Salamanca, I was suddenly struck straight in the midriff by a gale off that peak and sent hurtling out to sea.

Later Saint-Exupéry learned that the winds through which he had flown had been measured at 150 miles an hour at the surface of the ocean. The turbulence that formed as they tumbled over the Andes made them unnegotiable by any plane built in his day.

Airline pilots today fly much bigger craft, and they seek altitudes of 30,000 feet and higher to avoid the eddying air above mountain peaks. Occasionally, however, they are caught utterly off guard.

One cloudless December day in 1996, on a flight over the Colorado Rockies, an American Airlines McDonnell Douglas MD-80 unexpectedly encountered a patch of what pilots call clear-air turbulence. The plane dropped several yards, then stabilized. A seven-month-old baby flew out of his mother's arms; his head hit the overhead bin so hard it broke one of the reading lights. Eleven other people were also injured, though none seriously. In December 1997 a United Airlines flight en route to Honolulu from Japan hit a large patch of clear-air turbulence and fell nearly a thousand feet. One woman was killed; more than eighty other passengers were injured.

Clear-air turbulence is caused when currents converge to form what physicists call a chaotic state. If it were visible, it might resemble the eddies in a rocky stream, although it can take many different forms. It would pose a larger problem for commercial airlines but for the simple seat belt. Most travelers, even those who fly quite often, will never encounter clear-air turbulence so severe that the plane drops like an elevator. But it is more common in certain areas of the world, including the sky over the Rocky Mountains.

Eventually forecasters hope to be able to use lasers to detect clear-air turbulence far enough in advance to enable planes to steer around it. But traditional turbulence-warning devices, such as Doppler radar, depend on the presence of raindrops or other large, airborne particles and are useless for predicting rough air where there are no clouds.

Researchers are having better luck detecting the dangerous downdrafts, known to meteorologists as microbursts, that often accompany thunderstorms. When a plane flies through a microburst, it first encounters a strong headwind, then a chute of downward-flowing air, and finally a severe tailwind. Each sudden shearing of the wind can cause the plane to lose lift and even stall; if it is close to the ground it may crash. From 1964 through 1994 at least thirty mishaps

— some crashes, some near misses — were attributed to planes flying unexpectedly into wind shears. More than five hundred people were killed in the incidents. In the mid-1990s the Federal Aviation Administration began requiring commercial airlines to install wind-shear detectors in all their planes. Most airline companies chose to use wind speed sensors that alert the pilot once the plane has entered a dangerous wind field but that give no advance warning of wind shear. However, a few airlines have outfitted their planes with more sophisticated lasers and microwave radars that can detect wind shear up to forty seconds before the plane enters it, giving the pilot time to prepare for potential problems. The warning systems, designed by engineers at the National Aeronautics and Space Administration's Langley Research Center, are considered the first wave of a technological breakthrough that may make air travel considerably safer.

IT IS AN August day, and I am ensconced in a wicker chair on the porch of our rental cottage on a New Hampshire lake, where our family has been staying for a week. A freshening breeze crosses the water, promising a good, challenging sail — if only Jeff would come back from a trip to town.

On the floor beside me, our son leafs through a pile of books. "You haven't read to me all day," he complains. That's true; I haven't. I eye him and gauge my chances at making a deal.

"I've got an idea," I say. "I promise I'll read you eight books — eight *long* books — tonight if you'll go for a sail with me now."

Reid frowns. He is only four and scared of sailing; the instability of the little Sunfish moored out front is more than he can take. None of my pleadings over the past seven days have coaxed him to set foot in the boat. Still, eight books is at least an hour's worth of reading. A big payoff.

"I won't let anything happen, I promise. I'm a very good sailor."

He frowns again, and I know I've lost him. "Can I watch a video?" he asks.

I follow the passage of a gust over the water's surface and sigh. All that wind going to waste! "What book would you like to read first?" I ask.

For people who enjoy sailing, having to work or stay indoors on windy days is a form of torture. I drum my fingers on the table impatiently; I pace the floor. I close my eyes and breathe deeply, trying in vain to relax. My mind drifts again and again to the same thought: it's getting late, and the breeze is bound to die soon.

I know many other sailors who suffer from the same affliction. Hank Helmen, a pilot for a commuter airline outside Charlottesville, Virginia, is the most wind-bewitched person I have ever encountered; he tracks the wind's speed and direction virtually every waking hour of every day. For many years Helmen lived in Norfolk, where he owns a small apartment building. On its roof is a wind turbine he built to generate electricity for the lights in the entrance and hallways.

Helmen began piloting sailplanes when he was sixteen, but now he much prefers windsurfing. Years ago, when he was still living in Norfolk, he put several hand-carved wooden propellers, or whirligigs, on poles in his yard to help him gauge whether the wind was strong enough for sailing. "I'd been skunked so many times," he says. "I'd listen to the local forecasts, and they'd tell me the wind wasn't going to be any good, and then it'd blow like hell. Or else they'd predict wind when there wasn't any." One whirligig was designed to begin spinning when the wind reached ten miles an hour, another when the wind reached twenty, and another when it reached thirty. "When it got to twenty, I'd get my board and take off," he says. For a while he built whirligigs of various designs to sell in local novelty shops, but he stopped when the work took too much time away from sailing.

Several times a month Helmen goes back to Norfolk to windsurf; he has also sailed in many foreign countries. On days when the wind is good but he can't get away from work or other obligations, "There's this sinking feeling that comes over me," he says. "I can't really explain it. It's almost like guilt."

As if the wind were going to waste?

"That's it exactly," he says.

"When you're out sailing," he says, "and you catch a gust and sheet in the sail, it's just like there's a big hand pushing you along. It's the same feeling as going up a hill on a motorcycle and banking into a curve and giving it some throttle. It's the same thrill." His favorite places to sail are the Columbia Gorge and the Hawaiian island of Maui. Both are meccas for windsurfers. In the Columbia Gorge, summertime westerly winds meet a strong ocean-bound current and set up wave swells of up to eight feet. Since the 1970s various beaches near the town of Hood River have been given names by windsurfers: Bozo Beach (which is where you end up if you're a beginner, incapable of tacking upwind), Fish Hatchery Beach, Doug's Beach (made famous by a local equipment shop), and many more. The gorge, Helmen says, is the summer windsurfing capital of the world. Sailboard shops even have computer terminals that display maps of the air pressure along the river, so sailors can decide for themselves, at any minute, where the wind is best.

The direction and strength of the wind greatly alter the shape and height of ocean breakers, too. One day I happened to run into an old acquaintance, an aging hippie. When I mentioned that I was writing about the wind, his face lit up. "There are whole tribes that follow wind around the world," he said.

"Really?" I asked. What natives could these be that I had not heard of them?

"Oh, you bet," said my friend. "They'll go to Fiji, Australia, Costa Rica, wherever the seasonal winds carve out the best waves." Slowly his meaning dawned on me. The sun-loving surfer tribe. Of course.

PERHAPS no group of adventurers is so enamored of wind, or so much at its mercy, as those who take to the sky in balloons.

One afternoon a few years ago I picked up a copy of the *New York Times* and read an article about several teams who were vying to be the first to pilot a balloon around the world nonstop. One team from Albuquerque hoped to fly at the very brink of outer space in a capsule they had designed themselves and built with meager funds.

The balloon, then known as the *Odyssey,* would travel in a zone of scant oxygen, in air pressures so low that they could cause a person's blood to boil within eight seconds. The team members were training not only as balloonists but as astronauts.

To complete a circumnavigation of the world, balloonists must fly at least 15,835 miles in a great circle around the earth, under rules set by the Fédération Aeronautique International, a worldwide balloon organization based in Paris. Although they can adjust their altitude to catch the most favorable currents, they fly at the whim of the wind. The task of building a balloon that will stay aloft for the two to three weeks needed to make the trip is technically daunting. The propane required to keep a simple hot-air balloon up that long would weigh too much, and helium balloons that fly in the troposphere lose their lift as they cool at night. Most of the teams that have attempted around-the-world flights have used Rozier balloons, in which a bubble of helium is cushioned by a bed of hot air. A burner fired by propane occasionally reheats the air cushion and, in turn, the helium.

In the late 1980s and early 1990s, a Nevada-based team tried five times unsuccessfully to make the circumnavigation in a complex, double-tiered balloon named *Earthwinds.* The efforts reportedly cost between $7 million and $10 million. In January 1996 Chicago investor Steve Fossett managed to fly only from South Dakota to Nova Scotia in a simple $300,000 balloon before equipment failures forced him to land. Fossett flew from St. Louis to India in January 1997 and from St. Louis to Russia in January 1998, landing both times because of equipment failures.

A few weeks after Fossett's third flight a Swiss team took off from a site in the Alps and was forced to land in Burma when they could not get permission from Chinese authorities to ride the subpolar jet stream across that country. Other teams suffered spectacular disasters in the winter of 1997–98. A balloon carrying two Americans developed a tear shortly after its takeoff from New Mexico and had to be abandoned. A solo pilot who took off from Indiana aborted his flight when the base of his balloon burst. In Morocco, a balloon that was to have carried a British team was caught by a strong gust and swept aloft by itself, leaving the crew and capsule on the ground.

The failed attempts increased doubts among many pilots that conventional balloons could withstand the cold and stress of high altitudes, especially if teams seek to fly within the jet streams to gain speed and avoid storms. But the kind of balloon that was to be flown by the *Odyssey* team was of a design that had been used in high altitudes by atmospheric researchers since the 1950s. Dozens of such craft are launched each year, though not with humans in tow. In 1973 a scientific balloon took off from Australia, flew around the world twice, and landed only nine miles from its launch site. "There's no doubt in my mind that these balloons can stay up long enough to go around the world," says Steve Shope, the chief engineer for the project. "With the other balloon designs, there are questions about whether the envelopes will get too brittle in the cold and whether they can withstand the wind shears in the jet streams."

The winds of the stratosphere, where research balloons are flown, are among the smoothest and most predictable in the world because they are largely unruffled by weather fronts and by the earth's terrain. (They are also among the least studied and least understood.) Unlike jet streams, stratospheric winds spread out in wide, smooth flows, like a shallow river that spills across a plain. They circle the globe in gentle, undulating waves.

In summer the stratospheric winds flow from east to west, the opposite of the midlatitude westerlies of the troposphere. Sometime in spring the easterly winds lighten, and for several weeks the flow of air is fickle. In winter the winds reverse and blow strongly and less predictably from the west.

High-altitude research balloons are made of thin, clear polyethylene; they resemble gigantic sandwich bags. The type the balloonists planned to use is 900 feet in height and 400 feet in diameter and carries a gondola strung like a spiderling below it. For launch it is stretched out on the ground and filled with a small amount of helium while the gondola dangles beside it, held by a crane. Once released, the balloon rises quickly through the cold troposphere and the tropopause, the zone where clouds top out. Its size enables the helium to continue expanding, so it climbs to much higher altitudes than Rozier balloons. In the stratosphere it rides as high as 130,000 feet in the daytime and drops to about 80,000 feet in the cool of night.

Above the balloon is only the thinnest shell of the earth's outer atmosphere. The air is ringingly clear, the sky a velvety purple, even at noon. Below, the surface of the earth curves away, a cloud-shrouded plain.

The *Odyssey* team was to consist of either two or three pilots, including a television news reporter named Bob Martin, who had conceived the idea of using a research balloon for a manned around-the-world flight. Martin had begun searching for donors to fund the flight in 1993. He had unbounded enthusiasm, and he recruited dozens of volunteers, including several engineers who helped him design a capsule. Contributors gave about $1 million in the form of materials — computers, flight chairs, solar panels, and more. But *Odyssey* was perpetually strapped for cash. In 1995 Dymocks Booksellers, a large chain based in Australia, agreed to donate $500,000. The money was still not enough. Finally in the spring of 1998 Dave Liniger, chairman of the real estate firm RE/Max International, agreed to serve as the project's principal sponsor. Martin changed the name of the balloon to *Team RE/Max.* Liniger began training to fly on the mission, along with Martin and an Australian pilot, a champion balloonist named John Wallington. A firm launch date was set for December 1998 from Alice Springs, Australia.

"THE THINGS that worry me about this project," Steve Shope says, "are the mundane. Can we keep the temperature controlled in the capsule? Can we launch it without breaking the solar panels on top?"

Our conversation is punctuated by loud blasts from the propane burner that is keeping Shope's star-and-moon-spangled hot-air balloon aloft. We are 1,500 feet above west Albuquerque, floating in an urban haze. Below us is a dull patchwork of buildings. To the east the blocky Sandia Mountains hide the ascending sun. The best view is up, to where a nylon bag decorated with stars and swaths of color surrounds a globe of fire-warmed air.

The wicker basket creaks as the balloon rises and encounters a faster flow of air. "What I do is work my way up until I find some wind that I like," Shope says. "If you feel breeze on your face, it means we've hit a different wind layer and the balloon hasn't quite caught up with the flow." Indeed, I am surprised by how smooth the ride is, how dreamlike. "Wait till we land," Shope says.

While Bob Martin is the driving force behind the *Team RE/Max*, Shope, a physicist who runs an Albuquerque engineering firm, is the point man on safety. "One of the things you have to keep in mind is that until very recently this whole project has been built in a garage with a lot of ingenuity and volunteer help," he says. "Every real milestone in aviation has been accomplished by people working on their own. Look at the balloons built in France in the 1700s. Look at the Wright brothers."

The propane burner blasts again; we float in a placid sea of air. Below, a jackrabbit the size of an ant hops across a dull pink field. How different it would be to fly in a metal cocoon, insulated from the killing cold of space. At night the *Team RE/Max* will encounter temperatures as low as minus 150 degrees Fahrenheit; yet in the daytime the sun's rays could warm the capsule's air to lethal levels. To compensate, the gondola will be wrapped in a layer of foam and painted a shade of white that reflects the maximum amount of heat.

"We'll be putting the capsule through tests in a vacuum chamber," Shope says, "and we'll be able to adjust for temperature fluctuations then. We're going over everything in it — everything — with a lot of care. The balloon itself I'm not worried about. They've launched hundreds of balloons like this."

"See that field down there?" he asks a minute later. "That's where we'll land. Hold on and keep your weight low." And we come down with the precision of a small plane as jackrabbits scatter beneath us. The gondola bounces three times as we crouch, laughing; it tips and rights, the gorgeous balloon billowing above.

A short time later I enter a warehouse at EG&G, an aerospace contracting company that is building the interior structure of the Dymocks gondola for free. I expect to find something high-tech, but what greets me is an unfinished aluminum shell that looks very much like an oil tank tipped on its end. "You can see we've got a way to go," laughs Sabri Sansoy, a dark, friendly young man who serves as the team's structural engineer.

With its round portholes, the gondola reminds me of a flying saucer, though it is boxy instead of streamlined. Once it is aloft, the pilots will have to maintain a delicate thermal balance to complete their mission. If the helium expands too much, the balloon will rise too high, exposing the gondola to increased heat from solar radiation. But if too much helium is vented, the balloon may begin to fall. The danger of an unplanned descent will be greatest if the craft encounters a heavy cloud cover that shields the upper atmosphere from the earth's reflected heat.

Beneath a porthole the metal side bows inward slightly, as if something heavy had been leaned against it. "That shouldn't be a problem," Sansoy says. "We know the structure can withstand a 12-g shock, as long as all the welds hold. That's the stress it'll be under if the pilots have to abort the flight and open the parachute. There'll be a five-second free-fall, then it'll feel like they've hit a brick wall."

I climb inside. A mockup of the gondola, standing nearby, has a cheery white interior, with wide seats and desks and an elevated bunk. But the unfinished interior of the real thing is dark, echoing, and cold as a cave. When I stand in the center and open my arms wide, my hands come within eighteen inches of the walls. The ceiling, a mere five feet nine inches high, gives me a shiver of claustrophobia. Three men, plus supplies and equipment, for nearly three weeks.

The following day, flying home at 31,000 feet and 500 miles an hour, I watch the terrain beneath me, the brown crust crinkled into mesas and mountains, the houses glinting like silver beads. The earth seems fantastic and remote. I try to imagine being four times higher in a craft that flies as silently as a bird, moving so slowly the earth seems almost stationary below. Martin, Wallington, and Liniger are pioneers of a sort, striving to push back the limits of how humans can harness the wind. For the hundredth time I hope the men survive their jaunt in their homespun spaceship. Survive, and succeed.

I AM standing between two loaf-shaped California hills, listening to a symphony of wind. The air reverberates with mechanical hums, cricketlike squeaks, and a single, sustained note that sounds like a primitive, high-pitched flute. Mixed in are the melodious calls of meadowlarks and the buzz of blackbirds. But it is the industrial song, the sound of wind being put to work, that dominates the day.

I have come to Altamont Pass, northeast of San Francisco, to see and hear electricity being made. Altamont is the site of the biggest wind energy park in the world, with nearly 6,000 mills. Although many of the turbines here are outmoded, their sleek shapes — studding the hillsides, turning like giant pinwheels — form enough of a spectacle that traffic on Interstate 580 slows when they come into view.

Squeak, hum, squeak, whine. The turbines slice rapidly at the air, absorbing its flow. Each uses a rotating shaft and a generator to convert the wind-borne energy to electric current. Today's breeze is about fifteen miles an hour, brisk enough to chill me, but not loud enough to mask the turbines' sound. Their bright, white, tapered blades, so stark against the gently curving hills, are as beautiful as sculptures.

At the close of the twentieth century, wind power is considered the most promising source of renewable energy, largely because its cost has dropped almost by half since the late 1980s. Industry analysts say wind could provide 10 percent of the country's demand for electricity, and perhaps up to 40 percent if ways can be found to store its power efficiently. A few companies claim to be able to manufacture electricity from wind for four to five cents per kilowatt, a drop of between 80 and 90 percent from prices in the early 1980s. (This figure includes a one-and-a-half-cent federal tax credit, put in place in 1993 by Congress.) The turbines have been greatly improved, in terms of both their generating capacity and their safety and strength.

Yet most utility companies still regard wind power as too unreliable and expensive to ever become more than a minor, auxiliary source of electricity. As a result, it is considered a risky investment. In 1996 the Kenetech Windpower Corporation, the largest producer of turbines in the world, declared bankruptcy, casting further doubt on the industry's viability. (People familiar with Kenetech's operations say that the company experienced major technical problems with the blades and generator on a late-model turbine and that it suffered from questionable bidding and pricing policies.) With the widespread deregulation of the market for electricity, many people predict that the cost of electricity generated by fossil fuels will drop substantially and that the marginalization of wind power will increase.

Squeak, whoosh, squeak, thrum. Above me the turbines spin, driven by the same air that fills my lungs. What music, I think. What simplicity, what ingenuity. And what a shame that such brilliant technology is tied to an industry that is so political and so resistant to change. The generation of electricity has a long history of government oversight and labyrinthine regulations. It is also the most capital-intensive industry in the world, except perhaps for war. Not long ago advocates of renewable energy hoped wind power would give individuals and small collectives a way of meeting their own electricity needs, in a kind of democratization of the industry. And a few American companies do make turbines for users who need ten kilowatts of power or less. But the trend in the United States is toward building large, expensive turbines capable of generating hundreds of kilowatts — turbines so big that only major producers of electricity can afford to purchase them — and erecting them in huge wind farms.

Is wind ever likely to generate more than a tiny percentage of the country's, or the world's, electricity? The answer, it seems, depends largely on the rules by which electricity producers are forced to play.

THE FIRST electricity-producing turbines in America were erected in the late 1880s on the Massachusetts coast, and in 1890 a wealthy

Cleveland resident built a sixty-foot-tall turbine to light his mansion. But the first person to undertake extensive experiments on turbine design was Poul la Cour, a Dane who oversaw the operation of several turbines on the Danish coast after 1891. He became known as his country's Thomas Edison.

By the 1930s, hundreds of thousands of simple turbines (known then as home light plants) were in use on the American Great Plains, where the vast majority of farms were out of reach of electric utilities. The early turbines, connected to direct-current generators, were modeled after traditional farm windmills, with wide fan blades set at a slight angle to the wind. But as aeronautical science became more sophisticated, the blades on American turbines were reshaped to resemble airplane propellers and helicopter rotors. Through the 1930s wind power was widely used in rural areas to recharge batteries for vacuum-tube radios, the main source of news and entertainment.

In 1941 a Vermont man, Palmer Cosslett Putnam, built a turbine 110 feet high on a mountainside known as Grandpa's Knob. The turbine had stainless steel blades, each seventy feet long, and was designed to provide power directly to a local utility, a first. It was said to be capable of producing 1,250 kilowatts of electricity, enough to light a small town. Putnam's turbine worked for a year and a half, but in 1943 a bearing blew out, and a replacement could not be found until the close of World War II. Repairs were finally completed in February 1945. A month later one of the blades snapped off, and the turbine was shut down for good.

Small wind generators fell out of use as expanding utility companies and electric cooperatives hooked even remote households into central power grids. Electricity produced by oil, coal, or gas has a major advantage over wind power in that it can be generated on demand. Even in the most gale-lashed regions the wind does not always blow. Unless utilities can find a way of storing large quantities of electricity, wind power will always have to be supplemented by other electricity-generating facilities. (The same is true for solar power.)

Nonetheless, the notion of gleaning energy from a constantly renewable, nonpolluting source has long held tremendous appeal. In 1978, after embargoes by oil-producing countries in the Middle East drove up the price of oil and made consumers fearful of disruptions in its supply, Congress passed legislation requiring utilities to buy electricity from small, independent producers. The new law gave a tremendous boost to the development of renewable energy. In California, three areas had already been identified as potential sites for major wind energy parks: Altamont Pass, Tehachapi Pass east of Bakersfield, and San Gorgonio Pass west of Palm Springs. The state public utilities commission enacted a law requiring utilities to sign long-term contracts with renewable energy firms, guaranteeing the same price for each kilowatt of energy they produced as for power generated by fossil fuels. Then in the early 1980s, after Governor Jerry Brown began promoting the development of wind energy, California passed a package of lucrative tax incentives. Fledgling wind power companies began leasing land in the mountains and erecting turbines.

By concentrating development in a few mountain passes, the wind power industry shifted its emphasis from serving isolated on-site customers to feeding into the commercial power grid. Officials in the Department of Energy, assuming that the least expensive way to make wind power marketable would be to build turbines as large as possible, worked closely with the nation's aeronautical industry. Between 1974 and 1992 the Department of Energy spent nearly $500 million on research and development for wind power, much of it for huge turbines capable of producing up to 2.5 megawatts of electricity — more than three times the capacity of the largest machines that were actually in use in the late 1990s. (A megawatt is 1,000 kilowatts.) Without exception the large machines had major mechanical problems. None was ever produced on a commercial scale. "Some people compare it to the Wright brothers saying, 'We want to go flying. Let's build a 747,'" says Al Davies, a project manager with Enron Wind Corporation, a major manufacturer of turbines. In addition, the American reliance on aeronautical engineering was proving to be ill-placed.

In contrast, wind power companies in Denmark designed machines capable of generating 25 kilowatts, then expanded their size incrementally. In the 1980s they began widely marketing reliable, moderately priced 100-kilowatt turbines to California entrepreneurs. Instead of making light, ultraefficient blades modeled on aircraft propellers, the Danes found that heavier blades could better withstand the stresses imposed by wind. Airplane components operate for only a few hours at a time, while wind turbines must turn continually for thousands of hours. By adding heft to the blades and other turbine parts, the Danes designed machines with many fewer maintenance and safety problems.

Most of the turbines used in Denmark were erected in small clusters to serve farms, homes, and individual businesses. But the sale of turbines internationally, especially to the California companies that were building major wind parks, helped compensate for a decline in worldwide demand for Danish farm equipment. Even after the discovery of major oil reserves in the North Sea within Danish territorial waters, the country has continued to support the development of wind power.

Denmark's dominance of the world market is readily apparent at Altamont Pass. Most of the sleek machines that dot the hillsides are manufactured either by Danish companies or by American firms that have adopted Danish designs. German turbines, too, have become increasingly reliable. The German company Enercon is one of the largest producers of wind turbines in the world. By 1997 Germany had an estimated wind power production capacity of 1,800 megawatts, about the same as the United States.

The turbines now being built in this country were conceived from a blend of European and American philosophies: they are hefty and efficient but big. In 1996 the Zond Corporation unveiled a turbine 240 feet high, with a blade span of 150 feet and a generating capacity of 750 kilowatts. It is one of the largest turbines in the world. Based on its expected performance, Zond was awarded contracts to develop a 107-megawatt wind park on Buffalo Ridge in southwest Minnesota and a 112.5-megawatt park in northwestern Iowa. The Iowa park will be large enough to serve about 50,000 homes. (The

Minnesota project is the first phase of a 425-megawatt project, to be contracted to various companies and built by 2002.)

The fusion of American and European technology, along with the federal tax credit of one and a half cents per kilowatt, drastically lowered the cost of producing wind power and gave unprecedented momentum to the industry. In 1994 the National Renewable Energy Laboratory in Golden, Colorado, reported that 16,000 wind turbines had been installed in California and that wind power was producing enough electricity to meet the residential requirements of one million people. The industry was still plagued by a few persistent problems. For one, environmental groups had become concerned that wind turbines were killing large numbers of birds, including endangered falcons and eagles. For another, people who lived near large wind power plants often complained about their noise. But it seemed that both problems could be solved, as long as companies were careful to build new wind plants away from major flyways and residential developments. Longtime advocates of wind power were ebullient. It seemed that the technology was finally receiving the respect it deserved.

And then the bottom fell out.

In 1992 Congress deregulated the wholesale electricity market, enabling the largest users of electricity to purchase power directly from the cheapest suppliers. Natural gas had already been deregulated, and a worldwide glut had drastically driven down its price. (Between 1980 and 1996, deregulation and improvements in production technology allowed producers to drop the price of natural gas by 80 percent.) By 1996 industrial customers were pushing for the deregulation of the retail electric market as well.

With the industry in disarray, investment in wind power ceased. Renewable energy was still more expensive to produce than that from plants fired by fossil fuels, and as deregulation loomed, utilities began striving to lower their costs as much as possible. At the same time, the contracts in California that had long required utilities to purchase power from independent producers at fixed prices began to expire. Electricity generated by natural gas was selling for three cents per kilowatt or less, and utility companies were no longer willing to buy wind power for five cents. One by one, small wind power companies began downsizing or folding completely.

THE DAY before I visited Altamont Pass alone, I took a guided tour with Wayne Hoffman, a fiftyish man who knows firsthand about downsizing in the wind power industry. Hoffman joined Kenetech Windpower in 1987, then moved to FloWind Corporation in 1993 and was laid off in late 1995. A year and a half later FloWind is struggling to stay solvent. Hoffman now works in marketing and communications for Bechtel Corporation.

At Kenetech, Hoffman's first job was as an environmental planner at wind parks. Siting turbines can be a tricky business. When the speed of the wind doubles, both its force and the energy available from it increase by a factor of eight. A ridge with an average breeze of twenty-five miles an hour is much more productive than a ridge with an average breeze of twenty, and companies compete ferociously for leases on the most exposed hillsides. While turbines generally do not turn on until the wind speed reaches about ten miles an hour, many of them cut off at speeds over thirty to avoid being damaged. "Given all that," Hoffman told me over breakfast that morning, "there aren't as many good sites for turbines as you'd think."

Hoffman is a lanky, outspoken man with a clean-cut, boyish face and curly, graying hair. He described himself as a "wind wienie. That's what people like me are called by everyone else in the electricity business," he said. "Oil and gas are a whole lot more macho than wind power. Plants run by fossil fuels need pipelines and a lot of infrastructure that has to be built, and they need fuel to operate. That aspect appeals to people.

"Wind power gets into your blood," he added. "For those of us who got into this business for environmental reasons, it's been a difficult couple of years."

We drove east on Interstate 580 from overcast Berkeley, breaking into sun as we neared the velvety, treeless hills of Altamont Pass. The grassy terrain was losing its wintertime green and fading to gold in patches. On a distant ridge a row of willowy turbines appeared, looking at first like apparitions, like stick figures etched into the sky. Their blades turned, but slowly. "Not much wind out there today," Hoffman said.

As we traveled, Hoffman repeated several points I had heard from other advocates of wind power. It is misleading to talk about the low cost of generating electricity from fossil fuels, he said, because of the environmental price: pollution, acid rain, global warming. "If you considered all those things," he chuckled grimly, "you'd find you couldn't afford to buy any power made from fossil fuels. Then there's nuclear power." He did not have to finish the thought.

I asked Hoffman about a concept called the renewables portfolio standard, according to which electricity producers would be required to obtain a set percentage of their electricity from renewable sources. This would guarantee that the demand for renewable energy would continue to grow. The idea had won a degree of support in several states and in Congress. But the California legislature had recently refused to adopt a renewables portfolio standard, largely because of opposition from utilities in San Diego that had no sources of renewable power. Instead the state had adopted a "systems benefit charge," which will tack a few cents on to each electric bill to raise $540 million over four years for the development of renewable energy. "The RPS would have raised two to three times that much," Hoffman said, clearly disgusted.

"Look at what the fossil fuel industry is getting in terms of government support and tax breaks. And they don't pay any penalties for releasing greenhouse gases. It's not a level playing field." I had heard this argument before too. In addition to receiving generous allowances for depreciation on their equipment, power stations that burn fossil fuels do not pay tax on their fuel. Oil, gas, and coal companies often lease the right to mine public lands and waters at low cost. Nuclear power stations do not pay for insurance; the federal government underwrites them. The developers of wind power parks, on the other hand, are subject to all state and federal taxes, all liabilities.

We left the interstate and drove on back roads through the folded, shadowed hills. Here and there yellow wildflowers were smeared across the grassy slopes. I wondered where the turbines were. Except for three lines of tall machines on the westernmost ridge, the hillsides were bare. "I'm taking you this way for a reason," Hoffman said. "The view will be worth it." We rounded a curve, crested a hill, and stopped.

To the east and north stood several thousand turbines arranged across the hills in haphazard rows, some turning, some still. Some on latticework towers, some on simple concrete or metal masts. There were too many machines to regard any of them singly; they had become a forest. All their blades were pure white, but a number of machines had green or red brakes on their blade tips that would swing out horizontally if the rotors began spinning too quickly.

Lying as it does between the cool, marine climate of San Francisco and the warm, dry Central Valley, Altamont Pass is a place of strong west wind, especially in summer. But Hoffman and I had come on a spring day of light breeze. The turbines on the westernmost hilltops turned slowly, while those just behind them stood motionless. As we watched, machines scattered throughout the park began to turn, while their immediate neighbors remained stock-still. Their action or inaction seemed almost whimsical. We were witnessing the play of wind across the hilly terrain.

"You'd think the ones on the highest hills would be the first to turn on," Hoffman said, "but that's not always the way it works. The machines on the front ridge take most of the energy out of the wind. There's a lot of turbulence behind them."

We drove slowly through the park, with Hoffman pointing out turbines of various designs: American machines dating from the early 1980s, with thin, two-bladed rotors. Three-bladed Danish turbines bearing the names Vestas and Bonus, both industry leaders. Archaic German machines built in the early '80s, with wide-bladed rotors and tails that made them look like old farm windmills. The tails allow the turbines to swivel into the wind, but most modern machines use complex electronics to keep themselves properly aligned. Several of the old German machines gave off a rhythmic squeak, like a swing set in need of oil. "That's definitely some kind of internal problem," Hoffman said. "The newer machines are a lot quieter."

As we wound our way among the hills I caught glimpses of eight vultures, a soaring red-shouldered hawk, and a red-tailed hawk perched on a fence post. The sight of them made me cringe, for the Altamont wind park is notorious as a killer of birds. In 1992 a study by the California Energy Commission concluded that as many as thirty-nine golden eagles were being killed at the pass every year, either by flying through turbines or by perching on nearby power lines. The conclusions caused several environmental groups, including the powerful state chapter of the Sierra Club, to oppose further development of wind power in the state. Later the authors of the study conceded that it contained a number of far-reaching assumptions that may have biased their results. Nonetheless, even the most outspoken proponents of wind power admit that care should be taken to avoid building wind energy parks along migration routes or in regions with large concentrations of soaring birds.

On a distant hillside I spotted two machines with blades shaped like large, elliptical hoops joined to a central shaft. These were eggbeater, or Darrius, turbines, a design used in the 1980s that had been mostly abandoned because of problems with blade strength. "Watch 'em turn," Hoffman said. "They're mesmerizing." The movement of each blade reminded me of a long, swinging jump rope standing on end. They were hypnotic to watch, true, but the shafts that held the machines upright were covered with rust, as if they had not been well maintained. Financial constraints had prevented many of the companies with turbines at Altamont from keeping them in good repair.

It was time to return to Berkeley. As we rejoined the interstate, Hoffman turned for a last look at the turbines.

"You'd be surprised how many people think they're ugly," he said. "The aesthetic issue is one of the big problems in the industry. People don't want turbines interfering with their view of the mountains." He shrugged. "But I guess beauty is in the eye of the beholder."

Hoffman's words come back to me now as I shed my jacket, get back into my rental car, and leave Altamont Pass, driving south toward the southern California town of Tehachapi and a meeting with executives at Enron Wind Corporation. Along Interstate 5 I encounter a line of huge latticework stanchions that march south, mile after mile, holding aloft the wires that carry electricity to distant towns. Wind power parks may be visually obtrusive, but so are transmission lines. Strict energy-saving regulations could drastically reduce the need for such ugly structures and also curb air and water pollution. But an aggressive energy conservation program is not high on the national agenda just now. I drive south, the heat of the Central Valley sun forming beads of sweat on my arms.

"THE BOTTOM LINE," says Robert Gates, "is that we build what we can finance. We invest in large machines with upwind yaws and a certain blade design and a large generating capacity, because we know this configuration can work."

Gates is senior vice president for business development at Enron Wind, formerly the Zond Corporation, one of the biggest wind power firms in the world. The company has recently been purchased by a natural gas producer based in Texas. We are seated in Gates's office, which overlooks the dry, chaparral-covered slopes of the Tehachapi Mountains. On the other side of the building, up the dry mountain slopes, is a wind power plant with some of the largest, newest turbines in the world.

Gates and I are discussing how the demand for electricity differs country by country. In parts of Europe, where open land is scarcer than in the United States (and where governments provide generous subsidies to wind power producers), electric cooperatives have begun installing clusters of ten to fifteen turbines to serve customers in the immediate area. In developing countries the market is for small turbines of perhaps 100 kilowatts to supply electricity to isolated villages. Mongolian nomads carry collapsible turbines with a blade spread of six feet from camp to camp; on windy days these provide a power capacity of between 50 and 200 watts.

In the United States, Gates says, "the market is much more price sensitive. Everything is driven by cost. So you have to work with economies of scale. That means working with big machines. There's no getting around it." The major costs in the wind power industry are in capital equipment, since the machines do not need fuel. To make money, companies must use turbines that are both efficient to operate and strong enough to withstand the stress of high turbulence and blowing dust for many years.

I ask Gates about the wind plants that Enron Wind will soon erect on Buffalo Ridge in Minnesota and in northwest Iowa. Why did the company win out over other bidders?

"Cost," he says simply. "Northern States Power in Minnesota set up the bid scoring criterion, and 60 percent of it was weighted toward price."

"What price did you come up with?"

"Between three and three and a half cents a kilowatt. On average."

For a moment I am speechless. The lowest price I have ever heard for wind power is four cents, even with the one-and-a-half-cent federal tax credit. "Are you confident you can make that price?" I ask.

"Yes. There's a lot of wind in Minnesota — it's a good resource, an average wind speed of seventeen miles an hour — so we feel that the cost of energy should be lower than in many areas. We'll still be making ample profit with three to three and a half cents." I hope he is right. The 750-kilowatt turbines that will be installed at the company's plants in Minnesota and Iowa have just come on the market. I do not like to think of what the future holds for wind power if the projects fail.

It is time to take a tour of Enron Wind's facilities in the surrounding mountains. I thank Gates and go to the lobby, where I meet Jean-Pierre Bourgeacq, a director of project development. I get into a company van with Bourgeacq and his colleague, Rafael Alcade-Navarro. With us are two potential customers who work for a firm in Ontario.

We wind up the dry hillsides and through a field of turbines known as Victory Garden, which Bourgeacq says was the first American wind power park to be connected to a commercial power grid. I find this difficult to believe, since Zond was formed only in 1981, but Bourgeacq assures me it is true. I remember a comment made by Wayne Hoffman: not many technologies have developed so rapidly with so few problems. Three-bladed turbines surround us in every direction, most of them perched on latticework towers. For the most part they are small, 100-kilowatt machines, built in the 1980s. A handful of the newest turbines could outproduce all of them, Bourgeacq says.

Climbing higher, we enter a field crowded with turbines bearing the logo of Vestas, all built in the early 1990s. Atop a ridge Alcade-Navarro stops the van, and we get out. To the east is a hazy valley spottily covered with rounded shrubs. On some hillsides the soil is a dull red, on others an odd green. A single peak of barren red rock juts from the middle of the valley. We are on the edge of the Mojave Desert.

And we are nearly at eye level with the tops of dozens of wind turbines, which give off a loud *wop-wop-wop,* a ceaseless racket. Their sheer numbers — are there thousands here, or merely hundreds? — overwhelm my sense of perspective. Everywhere I look, blades rotate furiously; it's like being in the midst of an army of planes. It occurs to me that I would not like to live near a wind power plant.

As the others get back into the van, Alcade-Navarro stops me. "Look," he says, pointing to the west, to the basin that channels the wind over this pass. It is wide and dry and gently sloping, with nothing tall to block the wind — not buildings, not rocks, not trees. "You can see how the wind moves through here. It's like a river, flowing this way. It even keeps trees from growing in the valley."

"Yes," I say, "it's just like a river."

We drive a half-mile south through a crowded field of turbines to the base of a model called the Z-40. Unlike most of the older machines, the Z-40 is set on a tubular metal mast. I stand beside it, listening to the low hydraulic hum of its engine, the quiet *swish-swish-swish* of its blades knifing the air. It is one of the most sophisticated turbines in the world, capable of generating 500 kilowatts at a time. A few hundred feet away stands its offspring and successor, the 750-kilowatt Z-46. Both machines are so carefully shaped as to be elegant; both have electronically controlled blades that can swivel or "feather" into the wind like sails to catch more or less power, as conditions dictate. But the Z-40 is a constant-speed turbine that must absorb the stress of wind gusts. The Z-46 is capable of turning at variable speeds, depending on the wind. Because of this, it is more adept at handling gusts, and so produces power that is more consistent in strength and more highly valued by utilities. There is also evidence that larger, variable-speed turbines are apt to kill fewer birds.

The Z-46 is not turning just now. A few yards from its base lie two fiberglass blades from another Z-46, bright white and impossibly long, like the bones of a whale. Each weighs more than two tons. I run my hand down one, admiring its shape. This is the future of wind power. Is it possible that the problems of noise, visual pollution, and the killing of birds can be solved if we build turbines large enough? I am not easily seduced by promises of technological miracles, and the claims of the people at Enron Wind seem too good to be true. But, oh, how I would like to believe them.

I am due back in Berkeley tonight, so when Bourgeacq and Alcade-Navarro take us down the mountainside to headquarters, I bid a quick goodbye. Driving north, I pass again the columns of power stanchions, the groves of nut trees, the yellow, folded hills that block from my sight the falling sun. I wonder what the state of wind power will be in twenty years, and hope I will be able to purchase it for my island home. Just as dark closes in, I reach Altamont Pass, where the turbines whir endlessly, endlessly.

TEN

Taunting the Wind

ON A BLUSTERY SPRING DAY a current of west wind passes low over an unsown farm field in Nebraska, picking up grains of precious soil and spiriting them away, to be thrown the next morning against buildings in downtown Iowa City. Now the breeze flows around the smokestacks of coal-fired plants in the Ohio Valley, gathering a cargo of sulfates and ferrying them on, dispersing them to eastern forests, eastern lakes, cities, and skies. It swoops to the edge of the continent, to Philadelphia and New York, where it whistles through concrete canyons, swirling trash in gutters, toying with pedestrians, and activating the special dampers built into skyscraper frames to minimize their sway.

Wind has blown in the same basic patterns since the formation of the earth some four billion years ago. For most of human existence, we have modified our lives to accommodate its whims. We have built boats designed to take advantage of its moods and used it to pump water and saw wood. We have incorporated wind into our religions. We have sewn prayer flags and let them fly.

Wind has also shaped our dwellings and settlements. The Pantheon in Rome has a ventilation hole at the peak of its dome so that air is drawn up and out, creating a perpetual draft. In the first century A.D. the Roman architect Vitruvius drew a plan for a walled city with streets and alleyways laid out in concentric octagonals to shelter pedestrians from winds that were then considered unhealthy.

Urban residents in hot, dusty Pakistan built houses topped with periscope-shaped scoops to capture wind and funnel it inside. In Iran another ingenious design employed a hexagonal tower with small windows or slits near its top. At night, when the wind was still, air flowed up and out the windows. But in the daytime the wind would enter the tower, travel down to the basement of the house, pass over moistened walls or perhaps a pool, and flow coolly through the living quarters. Such systems were first put into use at least a thousand, and perhaps as long as five thousand, years ago. In places they are used to this day.

Settlers on the windy American prairie built low-slung cabins of sod or dug caves into the eastern sides of hills for protection from the prevailing westerlies. When a family amassed enough wealth to build a frame house, they surrounded it with plantings of trees positioned to break the prairie gales.

In southern Australia, in the desert hamlets of White Cliffs and Coober Pedy, opal miners abandoned the hot, dusty tents and barracks they had been given and carved houses out of the soft sandstone and clay of surrounding cliffs. The practice of building dugouts, as the dwellings are called, began in the early years of the twentieth century. Interiors were arranged with long hallways and ventilation shafts that drew air through them whenever the wind blew.

We have turned our backs on so much. We seem intent now on erecting structures of wood and concrete and steel that spite the wind, that issue a constant dare: try and knock this one down, and this one. We are bent on building cities on the edges of storm-plagued oceans, fire-plagued deserts. We release pollutants with a calculated faith that wind will disperse them.

Every so often catastrophe reminds us of our arrogance and vulnerability. Of the many examples I could cite, one seems especially apt because it was based utterly on human foible. There were no great tempests to blame, no demonic storms. There was only a muscular winter wind.

In 1968 construction began on what was to be an urban work of art, the sixty-story John Hancock Tower in downtown Boston. The building's mirrored façade contained more than 10,000 windows. But Boston, perched on the northwest Atlantic, is in a region of robust winds. In 1973, as the tower neared completion, windows began to pop out, sending pellets of glass in showers to the city streets.

Extensive tests later revealed that the layers of glass and chrome in the windows were improperly bonded. But the falling windows alerted engineers to another, more important problem. The tower flexed and twisted so much in the wind that it was in danger of toppling over. To steady it, engineers installed a tuned mass damper, consisting of two 600,000-pound blocks of lead, on the fifty-eighth floor. They also replaced the windows and reinforced the steel frame with 300 L-shaped beams. The price of the repairs was $15 million, a tenth of the building's total cost. This was the fine paid by various parties for the crime of taunting the wind.

THE EXCESSIVE swaying of the John Hancock Tower was not the first instance of the wind humbling architects and engineers, nor is it likely to be the last. During the nineteenth century, as iron was incorporated into bridge construction, a number of suspension bridges collapsed in Europe because they could not withstand the force of the wind. The worst disaster occurred in 1879, when the Tay Bridge failed in Scotland, killing seventy-five people. Nonetheless, the late nineteenth and early twentieth centuries were heady times for structural engineers. With the completion in 1937 of San Francisco's Golden Gate Bridge — a 4,200-foot span built in a region of earthquakes, high winds, and strong-running tides — it seemed that no engineering problem was too tricky to be resolved with patience and ingenuity.

In 1940 construction was begun on a 2,800-foot suspension bridge across the Tacoma Narrows, about thirty miles south of downtown Seattle. The bridge was unusually slender, only thirty-nine feet from edge to edge. Although a structural consultant had grave reservations about the design, a number of renowned engineers produced calculations to show that it would be both splendidly beautiful and perfectly safe.

"As soon as the Tacoma Narrows opened in July," writes Henry Petroski in his book *Engineers of Dreams*, "drivers noticed how flexible it was, the wave motion of its roadway bringing cars that were ahead . . . alternately into and out of view as the pavement rose and fell." The bridge became a local attraction, a free amusement ride. On November 7, the clamps holding one of the support cables snapped in a forty-mile-an-hour wind, and the roadway began a violent heaving. A few minutes later the bridge twisted apart and collapsed. The deck was simply too narrow; in a freshening breeze it had less tensile strength than a strand of spider silk.

That mistake would not be quickly repeated, at least not on so grand a scale. But wind engineering was still in its infancy. The only facilities available to measure wind loads on proposed buildings and bridges were aeronautical wind tunnels designed to test aircraft flying at high altitudes. In these tunnels the wind flows smoothly, with none of the turbulence of the atmospheric boundary layer. During the late 1950s a team at Colorado State University, inspired by a young engineer named Jack Cermak, built what it called a boundary-layer wind tunnel, the first test facility that could mimic the complex gusts and swirls stirred up by the passage of air over the earth. Then, in the early 1960s, Alan Davenport of the University of Western Ontario devised a method of using statistical analysis to figure wind loads on structures, which helped engineers interpret the air pressure measurements they had observed on models of skyscrapers.

On a summer afternoon in 1961, National League relief pitcher Stu Miller took the mound in San Francisco's Candlestick Park for the final inning of the season's All Star Game. The National League was ahead 3–2, but the American League had runners at first and second. As always, it was windy in Candlestick Park; dust swirled across the field, irritating the players' eyes, seeping into their mouths. Miller began a windup, only to be blown off the mound. "Balk!" called an ump, and the runners advanced. By the end of the game, players had committed seven errors that they blamed solely on the wind.

The problems with the stadium were supremely embarrassing, both to the San Francisco Giants and to the city as a whole. Beautifully situated on a peninsula extending into San Francisco Bay, Candlestick Park became known informally as the Cave of the Winds. In 1963 city leaders decided to explore the possibility of erecting some kind of shield. For advice they turned to a group of meteorologists in Palo Alto and to Jack Cermak's Fluid Dynamics and Diffusion Laboratory.

One spring morning I visited Jack Cermak at Colorado State. At seventy-one he was smooth-skinned and dapper, with an old-fashioned chivalry. He ushered me into the wind engineering facility with the quiet pleasure of a parent showing off a favorite child. Cermak had conducted wind tunnel tests at the university for forty years; indeed, he is known as a father of the science. It was here that he and several collaborators built a scale model of Candlestick Park and its surrounding terrain and discovered that the field would not have been so plagued by winds if it had been built a hundred yards to the north. "The prevailing winds split around Bayview Hill and come back together right at the park," he said. The engineers suggested several solutions, including the construction of a dome over half the stadium. City officials chose only to extend the upper tier of seats around the field to serve as a partial barrier. The park has never lost its reputation as the windiest stadium in the country. "That study really pointed out the need to look at wind effects before you build a structure," Cermak said. "It was also really the first time we had a chance to match the test results to the field results. It gave us a lot of confidence to go on."

Colorado State's wind testing facility is on the western edge of Fort Collins, at the base of the Rocky Mountains. As Cermak escorted me inside, the light falling from the cavernous ceiling was an odd yellow. Our voices echoed off the concrete floor of the warehouse-sized building. From a distant room I heard the monotonous hum of a fan.

Just inside the door was a display of models used in the most famous studies conducted at the school, including a Lucite model of the World Trade Center in New York. In 1964 Cermak and his colleagues showed that in a so-called design wind — a wind that would push the building's structural integrity to its limit — the top stories of the Trade Center towers would sway as much as twenty feet back and forth. Ten thousand viscoelastic dampers had to be placed inside the frame of each building to dissipate movement. It was a landmark study.

"This is the model we worked with to measure pressures for the design of the glazing and cladding," Cermak said. The slender towers were fitted with pressure taps — about a hundred, Cermak said — like tiny silver fuses hammered into their sides. A change in air pressure across a tap would cause an electric impulse to be sent to a computer. "Now look at this one," Cermak said, pointing to a model of the Sears Tower in Chicago. The building, tested in 1982, was crowded with taps, nearly twelve hundred of them. In the tiered corners of the model, the taps were as thick as buckshot. "You can see we've gotten a bit more precise," Cermak said.

At the end of a display table was a black-and-white photograph of a skier in a racing crouch. "Back in the mid-eighties the Olympic Ski Team got into one of our tunnels so we could check how their posture, their costumes, and so on would affect the amount of drag," Cermak said. "We found out that in the most ideal position, the skier's head is tucked too far down to be able to see ahead." He paused and grinned. "We had to modify things a bit."

Cermak next showed me a simple boundary-layer wind tunnel, which from the side looked like an oblong box six feet high and sixty feet long. It was set on stilts and had windows along its test section. Inside, a series of metal chains had been laid across the floor at carefully measured intervals. They reminded me of sand ripples on a beach. Near one end of the structure, more chains were laid in concentric circles and a six-inch-tall house sat on a turntable. As air moved across the chains, it would be bent into curls and waves to mimic the passage of wind over rough terrain.

This tunnel was a closed system; air did not enter from one end but flowed around an enclosed rectangle, of which the test section was one leg. In a closed-circuit tunnel, engineers can better control the flow of air at slow speeds and can set up thermal inversions, air stratifications, and other atmospheric quirks.

Wind tunnels can be powered by fans that blow air into them or suck air through them. Cermak strongly preferred sucking tunnels, and the eight structures in the lab had been built accordingly. "Blowing tunnels are very inefficient," he said, "because the fans create a lot of turbulence and you lose a lot of energy in straightening the flow of air." He escorted me to the crème de la crème, a closed-system meteorological tunnel constructed in 1963 and powered by a surplus B-39 aircraft propeller driven by an electric motor.

We threaded our way beneath thick white refrigeration pipes that kinked in all directions. Climbing a short set of steps, we reached a landing that overlooked the test chamber of the meteorological tunnel. It was ninety-eight feet long, with an inch-thick metal plate beneath its floor that could be heated or cooled electrically. Controls and gauges lined the wall behind us. "We can do just about anything in here that the atmosphere can do," Cermak said with obvious pride.

On a turntable sat an aeroelastic model of a five-story steel-frame building that was to be constructed in Taiwan. Cermak had been asked by the developer to test the stability of the building in high wind and during an earthquake. He was to compare how much the building would vibrate if built with a conventional steel frame and if fitted with special dampers to absorb movement.

He pointed to a stack of amplifiers, one of which had a small video screen. "Those are wired to sensors on the model," he said. "Watch that screen." Climbing into the test chamber through a small door, he walked softly toward the model and jarred it lightly. Immediately the flat green line on the monitor jumped, becoming a set of jagged fangs that smoothed into diminishing waves.

I joined him in the tunnel. The model, meticulously fashioned to scale, was pretty, with white canvas cladding, thick brass floors, and a bolted metal frame that reminded me of an Erector Set. I looked down the length of the test section, which was deceptively plain. Within this long, squat room engineers had conducted some of the most sophisticated wind studies in the world, creating gales, inversions, wind shears, and atmospheric stratifications.

WE LEFT the lab for a tour of Cermak Peterka Petersen, Inc., a consulting firm Cermak formed in 1981 with another engineer. On the way to the firm's offices, Cermak told me that when the first boundary-layer wind tunnel was completed at Colorado State in the early 1960s, many meteorologists doubted it could accurately model winds. "They didn't believe it could be done in *any* tunnel," Cermak said. "It took years to win them over." The computational fluid dynamics used to model wind flow have since become so sophisticated that engineers are now experimenting with numerical analyses of wind pressures, using computers.

Compared to the lab at Colorado State, CPP's facility was small and warmly lit. In the test section of a recirculating boundary-layer wind tunnel, Cermak's partner, Jon Peterka, crouched over a squat Styrofoam model of a new terminal building for the airport in Portland, Oregon. Again and again he inserted a gauged wire into piles of what looked like brown sawdust on the building's roof. The material was ground-up walnut shells, which, Cermak said, exactly mimicked the way snow would blow and drift. Nearby an assistant recorded the millimeter depths that Peterka called out.

On the other side of the building, workers painstakingly placed blocks in the test section of an open-ended tunnel to create the wind eddies that might plague a Titan IV missile awaiting launch at Cape Canaveral, Florida. NASA had asked the engineers to see how steadily a missile would stand in extreme winds. "They want to see what will happen if they have one ready for launch and a hurricane comes," Cermak said.

All around the walls of the room were shelves holding models from completed projects. The Disney Dolphin Hotel in Orlando. The Hawaii Prince Hotel in Waikiki. Office buildings in Bangkok, Manila, Seoul. As space grows increasingly scarce in urban areas, builders are forced to erect the tallest possible structures on tiny lots.

They could build skyscrapers that stretch upward for miles, if it weren't for earthquakes and wind.

CPP had also worked extensively on pollution dispersal, advising companies on such questions as how tall they must build smokestacks to keep plumes from settling too quickly and how best to vent fumes from parking garages and laboratory facilities. But that day I was most interested in the strange phenomenon of pedestrian-level winds, which can whistle through city canyons with brutal force. I once lived in Boston, and as I walked the downtown streets the wind assaulted me from every direction, viciously random and cuttingly cold. I would turn a corner and find myself walking into a gale, then turn another corner only to find the wind again in my face. How can that happen? I asked.

Cermak chuckled and took me over to the model of a glass skyscraper. "It's simple, though it probably doesn't seem that way," he said. "When you have a building sticking up above others, the wind hits it and travels down its face. You get extreme downdrafts at street level. You can alleviate that by putting the building on a pedestal a couple of stories high, so the wind hits the roof of the pedestal instead of the street.

"Now think about what happens when a strong wind passes by the edge of a building." He took his hand and ran it across the skyscraper's side, then crooked it quickly across the structure's front face. "The building exerts pressure on the wind. As the wind flows by the corner, its tendency is to get sucked around, across the side. There's a vacuum created." The air is pulled around the corner into the vacuum. "So pedestrians feel it from several different directions."

Some cities have enacted statutes requiring that new buildings be designed to keep pedestrian-level winds at manageable levels. "Hartford, Connecticut, has been one of the leaders in that," Cermak said. "Boston, too. We performed wind tunnel tests on the International Trade Center there. We had to rearrange those three buildings several times before the city would approve the pedestrian-level wind impact. It was quite a game of chess."

I walked with Cermak to the lobby of his office and thanked him for his time. He smiled graciously. "A lot of problems would be avoided if more people took into consideration the effects that wind can have before they make decisions."

IMAGINE STANDING in an empty house on the ocean as a major storm — say, a hurricane with winds up to 115 miles an hour — gears up to take it apart. Imagine the sweep of a powerful wind through a broken-out window and the back-and-forth flapping of a shattered door. The shudder of the frame as storm surf assaults the pilings. The creaks and snaps of the roof being peeled back. One afternoon I stood in an abandoned cottage in the Outer Banks town of Kitty Hawk while tendrils of a twenty-mile-an-hour west wind moaned through the living room, let in through an open door. A south-facing sun porch leaned precariously, as if ready to take the ten-foot plunge to solid ground. Above me, invisible behind a cosmetic sheath of cedar paneling, the roof had been framed using a simple toenail construction, each rafter held only by three nails pounded in at a slant. The house was like a million others along the beaches of the Atlantic, built with no more thought of hurricanes and northeast storms than if it had been a bungalow in Illinois. In a few days it would no longer exist.

It would no longer exist because that afternoon a team of researchers, armed with hydraulic jacks and stress gauges and electric saws, was scheduled to begin systematically taking it down.

Out the front door I could see the Atlantic, a dancing mosaic of sky-colored chips. The house shook as someone came up an outdoor staircase. David Rosowsky, a structural engineering professor from Clemson University, entered the living room, rolling his eyes. "We've got so much prep work to do, I don't think we're even going to get started today," he sighed. "We've got to take down the paneling, tear off the shingles, get rid of the insulation in the walls . . ." His voice trailed off. As a house is built in layers, so is it unbuilt.

Rosowsky was one of a handful of faculty members from Clemson working in conjunction with Blue Sky, a program that examined how coastal houses might be built, or rebuilt, to make them resistant to wind and flood. The project was based in the Outer Banks community of Southern Shores, where participants had built a prototype of a storm-worthy house. At Clemson scientists were studying the flow of wind around low-rise structures and designing machines to gauge the strength of windows and walls. All with the hope of finding ways to keep houses from breaking up during storms.

Until quite recently, structural engineers paid scant attention to the effects of wind on low-rise buildings, preferring to use their wind tunnels to test flow around models of urban skyscrapers. Wind tunnel work can be expensive, and no grant money was available to study the integrity of single-family homes. Besides, taller buildings are much more susceptible to strong upper-level winds than simple two- and three-story houses. But concern about the safety of private homes greatly increased after several devastating hurricanes — most notably Hugo in 1989 and Andrew in 1992. Also, after those storms, several major insurance companies canceled all homeowners' coverage in the southeast coastal plain.

In 1994, worried about the vulnerability of Outer Banks homes to storms, Cay Cross, then the town administrator for Southern Shores, attended a national hurricane conference in New Orleans. At the conference, she says, "it became clear to me that the technology exists to build houses that are much stronger than we're building them." Most dwellings in coastal areas were still being erected using standard construction techniques. A twenty-five-year lull in hurricanes had given homeowners the mistaken impression that such storms were rare. And since the 1950s the population of the Atlantic and Gulf coasts had more than doubled.

That fall Cross submitted a proposal to the Federal Emergency Management Agency (FEMA) for a program that would evaluate building codes in hurricane zones, teach building instructors basic wind engineering, and train builders to construct storm-worthy houses. She envisioned Blue Sky as a cooperative effort by government officials, engineers, builders, homeowners, and insurance companies. "If we can't build good houses while keeping costs reasonable, and if we can't come up with standards that the builders can meet, it ain't gonna fly," she said. FEMA agreed to contribute $1 million a year; another $1.14 million was to be raised from coastal states and private sponsors, including banks, roofers, lumber companies, and window and door manufacturers.

The primary goal of the project was to encourage homeowners to fortify existing houses or, in the case of new dwellings, to incorporate special construction techniques that would better withstand high winds. Engineers were experimenting with different materials and building methods that would, they said, make the price of a new, Blue Sky–approved home no more than 5 percent higher than that of a traditional home. It was an ambitious goal. Many of the recommendations they developed required the use of special materials or labor-intensive techniques. Notching rafters to fit snugly against the top plates of walls is effective but costly. So is reinforcing structures with galvanized steel rods. But the project was also pushing a number of small, inexpensive changes in construction practices. For example, if roof shingles are cut to extend just a half-inch over the edge of the roof, instead of the standard inch and a half, they are much less likely to blow off. Another part of the project involved designing and testing special steel clips and straps that could be nailed to roof and wall connections in existing houses, like metal Band-Aids.

Dave Rosowsky is a specialist in what engineers call resistance, a building's ability to hold together in a raking gale. He enjoys puzzling over why one building splinters apart in high wind and another does not. "I've spent a lot of time looking at human error in design and construction," he says.

In the spring of 1995 Rosowsky learned that Blue Sky owned a house in Southern Shores that was to be demolished. "I asked if we could come up and play around with it." The house, a single-story, wood-frame structure on a concrete slab, had been built about 1970, years before residential construction techniques had taken into consideration such factors as wind uplift, which is common in hurricanes. Rosowsky and his colleagues nicknamed it the Greenhouse for its pale green aluminum siding.

"We pushed and pulled and had a tremendous time," Rosowsky says with a slightly devilish grin. "The idea was to test our equipment to see how well we could measure resistance and determine the load paths. This is all new stuff; it's not too often that you get a house to tear down. At each step we were able to sit back and evaluate our results."

The critical factor in whether a house will hold together or break up in a storm is the distribution of the load placed on it by wind. In a well-built house the load is channeled through a pathway of sturdy supports to the foundation. The only way for the wind to destroy the house is to pick it up, foundation and all. In a more typical structure, the load is concentrated in one or two areas: the peak of a roof, a corner surrounded by windows. One weak link can cause the structural integrity of the house to fail.

Rosowsky and his coworkers applied steady pressure against the walls and roof of the Greenhouse with hydraulic jacks, mimicking the force exerted by wind as closely as possible. Over three days they rammed out windows and banged through doors. They tested the strength of roof–wall connections in one part of the house, then reinforced another section with hurricane straps and clips and tested those. They hired a crane to pull off a section of peaked roof—without success. "We thought it would be easy to get off; it didn't seem very well connected," Rosowsky says. "It wouldn't budge." Finally they sawed up the roof with the connections intact and trucked it back to Clemson for testing in the laboratory. "The nails in it were much bigger than what was required by the building code," Rosowsky says. "But it also turned out that the wood in the roof had dried just right, so the joints fit together incredibly tightly.

"The lesson learned is that it's not enough to test a house right after it's built. You really need to look at it over time to see how it strengthens or weakens in the field. Wood is an amazing material. It changes."

THE FOLLOWING DAY I returned to the abandoned oceanfront cottage to find it bustling with engineers and spectators, all wearing hard hats and safety glasses. The paneling on the living room ceiling was gone; so was the insulation and wiring and the shingles from the roof. Bright sunlight filtered through cracks in the plywood sheathing.

Along one wall was a line of shiny yellow ooze, dripping in thick stalactites. I touched it and was surprised at its firmness; it seemed like some kind of plastic glue. "That's FoamSeal," Rosowsky said, coming up behind me. "It's used to hold together mobile-home panels and to keep bumpers on cars; it's a really strong adhesive. We're hoping it'll be good for retrofitting houses, especially the ones built a long time ago, before there were decent codes."

On the edge of the living room a FoamSeal representative activated an applicator with tanks and pumps and hoses. As the pumps began to thrum, he climbed an A-frame ladder and squirted a milky-colored liquid in a line where the roof sheathing met one of the rafters. The instrument he held looked like a thick metal version of a child's squirt gun, connected to a fire hose. The FoamSeal quickly hardened and turned yellow. "It looks so simple to put on," I said.

"It is," Rosowsky said. "If it works for this kind of application, it'll be great."

I made my way to the house's enclosed underside, where a structural test was to be conducted. In the mid-1940s, when the cottage was built, it sat on a concrete slab on the ground. Later the owners had the structure set atop pilings and added the sun porch. Still later they built walls around the pilings to enclose a new ground-level story. But the beach had since eroded, and the ocean had started washing underneath. Recently the house had been condemned and given to Blue Sky.

I found several engineers in a room at the front of the house, inspecting a hydraulic jack. The device had been propped on its side with its shaft extending horizontally; it was positioned so it would push against a piling in the middle of the house and against an outside wall. Several long wooden braces had been nailed beneath a window so the pressure would be uniform across the wall.

Spencer Rogers, the researcher in charge of this test, was not from

Clemson but from a North Carolina coastal research program known as Sea Grant. His goal today was to duplicate the force that would be exerted by water rushing through the house. "A four-mile-an-hour current can apply the same pressure as a 110-mile-an-hour wind," he explained. "And that's just flat water. If you get a breaking five-foot wave, the force is greatly increased."

The jack rumbled to life. A gauge on its side relayed the pounds of pressure per square inch it exerted. "Eight hundred," Rogers called. "Nine-fifty. Twelve hundred." The braces began to bow outward. At fifteen hundred pounds wood began to crack.

There was a scraping sound from the middle of the room. An engineer standing by the jack glanced up to a beam that supported the top floor. "Whoa!" he shouted. The piling that braced the jack, and the beam, had canted several inches.

Rogers cut off the power on the jack. Engineers huddled below the beam, surveying the damage. "Time to come up with Plan B," someone said.

Rosowsky, standing off to the side, grinned widely. "Well," he said, "instead of busting out that wall, I guess we just tested the lateral strength of the interior piling."

"IT'S INTERESTING and a little frightening to realize how little was known about wind loads on low-rise buildings as recently as the early nineties," Scott Schiff said with a slight grimace.

Schiff, a professor in the department of civil engineering at Clemson University, coordinates the school's affiliation with Blue Sky. Until 1989 he worked at the University of Illinois and concentrated on finding ways to minimize the damage to buildings from earthquakes. That year he moved to Clemson to study wind loads and building resistance. He arrived just before Hurricane Hugo, which made landfall on the South Carolina coast and traveled inland across half the state.

After the storm Schiff and other Clemson faculty members convinced FEMA officials to fund a study on how low-rise buildings might be made less susceptible to wind damage. "It was a hard sell," he said. "No one had ever done anything like that; the feeling was, if you build a house in a hurricane zone you just have to take your chances." Eventually FEMA contributed $700,000 to Clemson for research on how winds — especially the multidirectional winds of hurricanes — place stress on buildings less than three stories high. With that money and $700,000 in matching funds, the university built a wind load test facility, the only such laboratory in the Southeast.

The facility included a boundary-layer wind tunnel, which was completed in 1992. "As soon as we got it up and running," Schiff said, "we sent three undergraduate students down to Folly Beach, on the coast, to take pictures of houses. When they came back we built models of the houses they'd seen and put them in the tunnel.

"We'd measure the loads on a house when it was sitting on the ground, and then we'd put it up eight feet on pilings and the loads would change. Sitting up high like that, it was hit by stronger winds. That didn't really surprise us, but we started asking, Why aren't insurance companies and building codes taking these kinds of things into consideration?"

The Clemson studies had been under way for nearly two years when Cay Cross submitted her grant proposal to FEMA. Blue Sky provided the engineers with both a new source of funding and a way to put nuts-and-bolts research to practical use. "The idea is to use the research to come up with better building codes," Schiff said. "Right now they're basically the same for houses in the mountains and houses in the coastal plain."

"Traditionally building codes have used safety as their goal; they wanted to make sure your house didn't fall down on you. We're taking it a step further. We want the house to go through a storm and still be habitable. Your carpet may be wet, your electricity may be off, but in the long run you'll still be able to live in it."

Schiff was hosting a corporate round table at the university for participants in Blue Sky. After our conversation I drove out to the Clemson Wind Load Test Facility, which was in a metal, hangarlike building down a shoulderless country road. In a room on the west side of the building was a blowing wind tunnel, powered by two huge fans. Air passed through four screens, a metal honeycomb, and another screen, all of which served to straighten it, then through a barrier of plywood spires to create riffles and eddies. Next it flowed over four planks propped on the floor at a 45-degree angle, over a checkerboard of three-inch cubes, and over a final grid of one-and-a-half-inch cubes to a turntable a hundred feet from the fans.

In 1987 engineers at Texas Tech University in Lubbock built a two-story metal building on the open plains so they could see whether the results of wind tunnel tests — results that had been compiled over more than a decade and were considered virtually irrefutable — held true when a real building was exposed to real wind. The building was constructed on a turntable, with pressure gauges mounted on its walls and eaves. By and large, the values predicted by the tunnel tests proved to be accurate: winds of *x* speed placed a pressure of *y* on the building. But where the roof met the building's corners, the engineers' predictions were disturbingly off. In subsequent experiments, researchers noticed small vortices spinning off the corners of buildings. A brisk breeze caused the formation of little tornadoes that doubled the pressure placed on the roof.

Blue Sky participants began drifting into the Clemson lab for a series of demonstrations. A student wearing a respirator climbed inside the wind tunnel with a long, thin stick, a giant Q-Tip that emitted a smoky white stream — titanium tetrachloride, a toxic substance formerly used in skywriting. The student touched the stick to a model of a ranch house as if he were painting it. Smoke streamed off the roof peak and spiraled off its corners, flowing in long, wavy streams, a cross between water and cloud. Wind made visible. At the very front of the house, near the ground, the smoke formed a long, rotating coil. "Notice the vortex at the front," someone called. "It's like a tornado turned on its side."

Indeed it was. Who would have thought a simple wind could execute such lovely twists and flips?

This was the most basic of wind tunnel tests. The fans were set at their lowest speed, about five miles an hour, to keep the smoke from dispersing. As the student climbed out of the tunnel, a faculty member turned up the fans, just for fun. A half-dozen of us positioned ourselves at the tunnel's end to let the wind pass through our hair. I caught the eye of an engineer who stood beside me, a longtime Outer Banks resident, and laughed. The fans were nearly at top speed, humming loudly, throwing off air at thirty-five miles an hour. To us, though, it was nothing special. "Feels like a typical spring day," the engineer chuckled.

The spectators moved outside to a concrete pad, where Scott Schiff and several students stood next to a model of an unfinished wall built with plywood sheathing and wooden studs. A beam at the top of the wall was bolted to a broad steel girder, against which a hydraulic jack had been placed. A student named David Stricklin explained that the jack pushing against the girder would mimic the force exerted on the wall by a wind howling below the eaves and threatening to peel back the roof. "We've tested forty walls like this," Stricklin said, "ten with openings — windows or doors — and the others without. Let's see what happens."

The jack sprang to life; the chains began to creak. But even when 1,200 pounds of load was being applied to each of the plywood panels, the wall remained solid. The student increased the load: 1,800 pounds, 2,000 pounds. The wood began to make cracking sounds and a single, startling POP. At 2,200 pounds the top beam pulled off, exposing ripped-out nails.

This was not what Stricklin had expected. "We predicted it would fail at the bottom," he said, "because we used southern pine for the top beam and fir, which is weaker, for the studs. I thought for sure the fir studs would fail first."

A house would probably never have to sustain such punishment. The pressure applied, Schiff said, was equivalent to a 180-mile-an-hour wind against a two-story oceanfront house elevated on pilings, with no open windows or doors. The wall had been built to code specifications that ensured it would survive a 110-mile-an-hour wind; the specifications included a large safety factor to guard against defects in building materials and workmanship. "That way,"

Schiff said, "if you get a 120- or 130-mile-an-hour wind, your house should still be okay."

The wall had not yet received its full share of abuse. Now two students nailed a second plywood sheet to it and readied a homemade "cannon," an odd contraption fashioned from an air compressor, a length of plastic pipe, and an empty beer keg. Pieces of wood could be shot out of the pipe at controlled speeds. With a loud *thwack* the cannon spit a two-by-four toward the wall. It slammed viciously through the plywood, a spear hurled at thirty-five miles an hour. "Debris thrown around by wind can do a lot of damage," Schiff said, a bit wryly.

I pulled on the two-by-four and tried to rock it free. It had penetrated both the plywood sheathing and the original wall; it would not budge. Most island residents I know worry about how well their own houses can hold up in storms. They seldom consider the damage that might result if their next-door neighbor's house were to fly apart and pummel them with debris. Another reason, I thought, to hope for a lifetime of moderate sailing and gentle seas.

IS IT REALLY possible to hurricane-proof every house along a coastline, as Blue Sky participants would like to do? Not without banning mobile homes, which are the most vulnerable of all structures to damage from wind. And the cost of retrofitting stick-built houses with hurricane ties and improved roofing is probably higher than many homeowners can afford.

One day shortly after my trip to Clemson, I happened to meet the Outer Banks engineer who worked as a consultant for Blue Sky. I asked him about the cost of incorporating some of the project's design recommendations into a traditional home. Was it realistic, I asked, to think it could be done for no more than 5 percent of the home's overall cost?

The builder chuckled. "It depends on who's building your house," he replied. "If you're putting up a no-frills beach box, and you've got one of these outfits where the guys can barely nail two sticks together, then no. But if you're paying for a quality job and you're using good materials, then yeah, I'd say it's possible. They're not there yet. But it's within reach."

Like everything else in life, you get what you pay for in a house. Satisfied with the engineer's answer, I went home and settled down with a cup of coffee and the morning paper. The local section featured a story about Hurricane Fran, which had ravaged the North Carolina coast a few weeks before. Scott Schiff and his team had inspected many of the houses that were left partially standing. The headline read "Fran damage blamed on shoddy building." It was a message I could no longer ignore. I left the paper on the kitchen table and went upstairs, flashlight in hand, to cast a suspicious eye on the connections between our roof and the walls.

ALTHOUGH it is the storm winds that set the adrenal gland pumping, that threaten life, limb, and the roof over one's head, it is the light winds of fair weather that most effectively bring to earth whatever poisons have been pumped into the sky.

In a small room deep within the Environmental Protection Agency's Fluid Modeling Facility, I watched as a technician aimed a laser at a square tank filled with water. For an hour and a half, tiny heaters had been warming layers of the water to precise temperatures, from 80 to 105 degrees Fahrenheit, to create an abnormally stable environment without currents, without even a whisper of bottom-to-surface flow. At the top of the tank a thick lid of warm water sat on the cooler strata, simulating the kind of temperature inversion that might form over, say, a Kansas prairie on a midsummer day.

Since 1974 investigators at the Fluid Modeling Facility in Research Triangle Park, North Carolina, have studied wind by studying water. Within both the square convection tank and a second tank more than eighty feet long, they can simulate inversions or windy turbulence or highly stratified layers of air like those that might form in the atmosphere on any given day. Today they planned to track the plume from a smokestack on one of the worst — that is, most windless — days of the year.

The scientists' findings are used to help set federal standards for the spewing of noxious plumes. "It's a good thing we have wind. Otherwise we wouldn't be able to release any pollutants at all," facility chief William Snyder told me.

A number of times the tests undertaken at the laboratory have led to the easing or tightening of federal standards on air pollution. In 1975 Snyder and a collaborator, Robert Lawson, Jr., conducted a surprising study that challenged a long-existing tenet of pollution control. A rural electric cooperative had applied for permits to build a lignite-burning plant. A second structure was to be built next to the lignite burner to house generating equipment. The adjacent building would block some of the wind, decreasing the speed with which smoke from the burner's stack would dissipate. In moderate to heavy wind some of the stack's emissions might even wash down to the ground. To eliminate potential problems, federal law stipulated that the smokestacks had to be two and a half times the height of the building, in this case 750 feet. A stack that high would cost an estimated $15 million.

"The officials from the company came to us and asked if they really had to build it that high," Snyder said. "So we ran some tests in our meteorological wind tunnel. We found that the building was thin enough that the blockage would be minimal and crosswinds would help disperse the plume." In the end the cooperative had to build the stack only 450 feet high, one and a half times the building's height, at a cost of $5 million. "Conventional wisdom turned out to be wrong," Snyder said. "A simple test saved consumers $10 million." Subsequent studies found that the lower smokestack ratio was sufficient in many cases. When Congress reauthorized the Clean Air Act in 1980, the requirements on smokestack height were amended accordingly.

Knowledge of wind flow patterns has since grown tremendously, and wind simulations were being conducted with more precision. Before my visit to the facility I had seen a video of a test in the convection tank, which measured four feet on each side and sixteen inches in depth. In the video a puddle of milk was heated so that droplets of it rose from the floor, stretching upward on thin necks like sprouting seeds and disappearing in smoky wisps. This was a demonstration of how air rises over "a heat island, a modern city," Snyder said, "something sizable, like New York."

The experiment I planned to watch was one of a long series of tests to measure the shape and dispersion of a buoyant plume, the kind produced by a coal fire. A model smokestack containing a liquid colored with fluorescent dye was to be towed slowly through the water in the tank to simulate the flow of wind. "It would be very difficult and expensive to build and heat a water-flow channel that could model convection," Snyder said, "so we move the stack instead of the water.

"The temperatures in the tank are controlled by computer; all we do is set up the program. Every time we run a test, even under the same conditions, we get a slightly different result. Turbulence is chaotic, so you have to do many tests of each situation to generate an average. We take a video picture of a cross section of the plume every second. From those we figure the most likely scenario in a given wind."

He handed me a pair of orange safety glasses and slid a pair over his own eyes. A video camera had been positioned in front of the tank to record the shape of the liquid plume. Beside the tank was a boxy case holding the laser that would illuminate the plume so it could be seen within the dark water and filmed for later analysis. "Ready?" Snyder asked.

A graduate student flipped a switch, and orange fluid began streaming from the stack.

"Watch carefully," Snyder said. "This kind of plume will penetrate into the stable layer and just sit there until some mixing occurs. Until the inversion breaks up. That could take a while, depending on atmospheric conditions."

I watched in awe. It was high-tech art, this bright orange well-spring, swirling and boiling through black water (black air), dancing and curling into strands, then abruptly hitting the temperature ceiling. Orange liquid spread in waves from the point of impact, thickening, hanging as if stunned, as if trapped — which of course it was.

On the other side of the room, a computer generated pictures of the plume shape. "It's amazing to me how quickly the cross section changes," Snyder said. Indeed, the images flashed on the screen with dizzying quickness. Now the plume was fat and rounded on top; now it had thinned; now it fattened again and curled toward the right. The profiles seemed to be taken by cameras placed at every conceivable angle, though only one camera was in the room.

"It's beautiful," I said, watching the plume in the tank widen and split into many silken eddies. In the silent moments that followed I wished deeply that such technical skill and poetry represented something besides a reaching, poisonous gas on a summer afternoon of light wind.

"WHO HAS SEEN the wind? Neither you nor I," Christina Rossetti writes in her most famous poem. But that is no longer quite true. Anyone who has noticed a dwindling of fish in the lakes of the Northeast, who has visited a grove of dying cedars on a mountaintop or seen the pockmarks in the gargoyles at the Cathedral of Notre-Dame has seen what wind transports, if not wind itself.

"All normal life stopped here simply because there was a strong [southerly] wind on April 26, 1986," writes Michael Specter in a special report to the *New York Times.* "People have become paralyzed with fear. They are afraid to move, afraid to marry, and afraid to have families." Specter is describing the Gomel region of Belarus, once an area of rich farmland, just north of the Chernobyl nuclear plant in Russia. Even ten years after the catastrophic accident at the plant, as much as a quarter of the land in Belarus was considered uninhabitable. Farms and houses stood with coats still in their closets, toys still strewn across their floors. Nearly 200,000 people had been resettled in other parts of the country, many of them to crowded apartments in the city of Gomel. With the dissolution of the Soviet Union, Belarus "was left with immense bills it could not possibly handle," Specter writes. "An agricultural land tainted by the ultimate modern poison is of little use to anyone." Residents are sick with cancer and various diseases, yes, but their biggest ailment may be fear. As I read Specter's article, it occurred to me that a century ago no one would have predicted that humankind could accomplish such a massive, instantaneous fouling of our nest. Who could have foreseen that the wind, the wonderful, cleansing wind, would one day transport such a deadly poison so effectively, so fast?

Much of the world's worst pollution occurs in areas where the wind scarcely blows because of temperature inversions and other atmospheric anomalies. Los Angeles, Denver, and Albuquerque, all lying at the bases of mountains, frequently generate a higher load of toxins than the wind can flush from their skies; residents must be constantly on the alert for yellow air. In other parts of the world, pollution warnings often accompany fog. In November 1953 a fog descended on New York City that carried a high concentration of industrial smoke. City health officials estimated that 250 people died, mostly of respiratory difficulties. That single incident killed nearly as many people as Hurricane Camille, the strongest tropical storm to hit the United States since the 1930s. In the modern industrial world, when the wind falls still, the consequences can be just as deadly as when it rages.

More typically, wind-borne pollution brings a slow disintegration of health, not just to people but to natural areas and the stone facades of city buildings. In the 1970s biologists noticed a marked decline in the populations of fish in lakes in eastern Canada and New England, and a decline in the lushness of high-altitude coniferous trees in Vermont and North Carolina. Similar problems were reported in Germany, Bavaria, Czechoslovakia, and the Scandinavian countries. At the same time, odd blemishes appeared on the facades of many of Europe's and America's most famous buildings. The cause was acid rain, brought by wind that passed first over coal-fired plants and metal smelters.

Acid rain is composed mostly of sulfur dioxide, which within hours of its release is converted into sulfate by-products that can be blown thousands of miles. In September 1988 wind and rain carried a potent dose of corrosive air eastward from plants and utilities in Atlanta, St. Louis, the Ohio Valley, and Sudbury, Ontario, home of the International Nickel Company. Scientists at the North Carolina

Supercomputing Center later created a computer simulation that tracked the spread of the sulfates. The simulation was incorporated into a video entitled *Caustic Sky*.

When sulfates are blown into the upper troposphere, they quickly encounter an atmospheric ceiling much like the temperature inversion layer I witnessed in Bill Snyder's convection tank. The troposphere is not a deep well of air but a shallow lake that tops out at roughly 40,000 feet. Above this is the tropopause, a stratum of dense air that marks the edge of the stratosphere. Pollutants generally do not penetrate into the tropopause. Instead they spread horizontally, trapped in the realm of weather, traveling eastward with the clouds over most of the United States.

In the computer simulation, sulfur dioxide emissions were represented by pinkish clouds. A brief section of the video showed these gases accumulating in piles, like drifted sand, around the factories where they had been released. The video next displayed a simulation of pure sulfur dioxide and sulfur by-products, represented by light blue clouds, being carried east. At first the sulfur dioxide gases streamed away from their point of release in pink smears, but they quickly turned blue as they mixed with the atmosphere and changed composition. The blue clouds spread, growing thinner, until they looked more like lenses of water. They moved eastward in a pulsing rhythm, chased from behind by a storm. Near the eastern edge of the continent they grew thinner still, and became huge bubbles that squeezed against each other as if fighting to escape the rain nibbling at their back sides. In the end most of the gases washed to the ground.

Medical professionals and environmentalists have long been concerned that the well-documented spread of wind-borne pollutants like acid rain represents only a portion of a much larger, worldwide problem. In late 1996 the Environmental Protection Agency proposed new restrictions on the release of small particulate matter and the class of ozone that is a major component of smog. The proposed restrictions were immediately criticized by industry representatives, both because of the expense that would be entailed in meeting them and because they would assign blame for downwind pollution to the locale where it was released. For the first time industries in the Ohio Valley, for example, were to be made responsible for the smoke and toxins they ship downwind to cities in the Northeast.

More and more, filth carried on the wind is a problem in the farthest reaches of the world. During the late 1980s, French and German biologists working in Africa discovered that acid rain was ravaging the rain forests of the Congo Basin, and acid fog was enshrouding much of Ivory Coast and Zaire. The announcement shocked natural scientists. Where was the poisonous air coming from? Later the French and German investigators concluded that it was being generated by the burning of millions of acres of savanna in western and central Africa each year to maintain the growth of grazing grasses. They held little hope that farmers and herders could be persuaded to forgo their annual fires.

IN 1990 Jeff and I took a trip to the ostensibly still pristine country of Costa Rica. Panama's northern neighbor is well known for its sound conservation policies, so we realized only gradually as we traveled the country that a great deal of rain forest had been cut to establish the pastoral farmlands. One day in Tilarán, a foothills town near the southwest edge of Lake Arenal, we met a young man named Carlos who had lived in the region all his life. The lake, a dammed reservoir, had recently been discovered by American and European windsurfers. We joked with Carlos about how the plain, working-class town would soon be overrun by windsurfers wearing fashionable clothes — if they could stand the constant upslope gales. The wind was ceaseless and brutal; to hold a conversation on the street we had to take refuge in doorways.

Carlos laughed at our reaction to the wind. "We traded the rain forest to get all this," he said.

"What do you mean?" I asked.

"It didn't use to be this windy at all," he replied. "It's happened in the last ten or twelve years, since so many of the trees were cut." Later I spoke with several of Tilarán's older residents, who confirmed Carlos's observations.

I know of no studies that carefully document the correlation between destruction of rain forest and shifts in regional patterns of wind and rain. But tropical biologists have witnessed the phenomenon described by Carlos in many countries. Without trees to slow it, the wind builds and howls across the land. In the process it erodes depleted soils and desiccates struggling crops. It carries rain clouds quickly through countryside where once they lazily drifted.

With our appetite for natural resources, our passion for travel, our insatiable quest for commodities, humans are changing patterns of atmospheric circulation in ways we can scarcely imagine. Consider the greenhouse effect, in which the burning of fossil fuels and the release of harmful gases may cause the temperature of the earth to rise. There is still no agreement on whether the earth is warming; instead, some scientists believe an increase in cloud cover over the oceans may serve as a thermostat to regulate the temperature of the planet. But what will happen to global wind patterns if predictions of a warming trend prove true?

The realm of the prevailing westerlies will weaken and shrink in diameter; the horse latitudes will drift toward the poles, enabling the great deserts of the world to expand. Monsoons will be strengthened and bring more rain, perhaps double the amount of present times. Hurricanes will form more frequently and perhaps amass greater power. Those, at least, are a few of the predictions of climatologists. Even the most thoughtfully conceived theories by the world's most talented scientists are no more than educated guesses. The only thing known for certain is that the winds of the world will change. Who of our ancestors could have predicted that our species might grow so omnipotent as to reshape the breath of God?

EPILOGUE

A World of Great Wind

IT IS LATE FALL, and I am driving toward home after a long time away. This morning I awoke to a quiet stirring in the Virginia Blue Ridge, a soft clattering of leaves that in no way resembles the deep, whooshing sighs of loblolly pines on the Outer Banks. That delicate, foreign sound made me realize with a pang how ready I was to be back in the landscape where I belong.

I drive with the windows partway down, waiting for the first scent of salt air. The grasses and trees along the road ripple in a steady breeze. The closer I get to the sea, the stronger the wind. Sixty-some miles from the Outer Banks I pass a creek thinly lined with *Juncus roemerianus,* black needle rush, a coastal species and the westernmost sign that I am entering the natural system of home. A few minutes past the creek I feel a sluggishness in my truck, and when I let my foot off the gas it quickly slows. I realize suddenly that this is no ordinary wind. A northeaster is building, cleansing the Outer Banks of sultry air and holding at bay any less-than-hardy souls who have no place there.

I pass quickly through the town of Edenton and drive south over Albemarle Sound. Silt-laden waves crest on the open water and slam into the bridge pilings. I have to fight to keep the truck from drifting toward the right — toward the west. The wind just now is like a huge door swinging shut, barring the way for creatures that have not already settled into a niche on the islands. If I can reach our sheltered parcel of maritime forest, I'll be safe. But the gale doesn't want me there; it's slamming me out with everthing else. "I belong on the banks!" I shout to thin air. "It's me; let me in!" It occurs to me that if I stay low, if I just push on, I might be able to slide beneath the crack in the door. Only someone who loves the wind would know to do that.

I LIVE in a land of great wind, and it defines me. Whenever possible I seek it out, walking the beaches, climbing high dunes to revel in its touch. Again and again I marvel at its changeable nature, its many personas. Nothing on earth so cleanses my mind or pushes me so near the brink of my physical limits. Nothing else so reminds me of God.

One day after my return from Virginia I awake with a terrible cold. My thoughts feel as if they are locked in a small, dark closet, and with each swallow my throat roars. I stay in bed an hour past the alarm, then with a start I remember that a crew of men are coming this morning to begin putting a new roof on the house. I had better find somewhere else to sleep.

Where can I go? Any number of friends might invite me into their homes, but I don't want company, and I would hate to spread this sickness. Retreating to the family car, I take with me a blanket, a pillow, a jar of ice water, and several books. I am lucky about the weather. It's a bright, sunny day, with a feisty but warm breeze. A small-craft advisory is forecast for this afternoon, but maybe the car will provide enough shelter. I drive to the north end of the island, to a sandy parking lot that overlooks a small, marshy pond and wide Croatan Sound.

In the shade of a pine I park the car, roll down the windows, and push back the seat. My head is on the pillow, my arm thrown over my eyes. For a long time I drift in and out of sleep. The wind builds and builds, until the gentle waves on the sound become two-foot foaming breakers, and the wax myrtle bushes creak and sigh, and the pines sing. The wind takes on a palpable presence, a being moving in and through the car, attending to my needs. It is not a dragon today, but a person. A man. What was that name I once heard? This wind is the harmattan, the doctor, the attending physician. It massages me; it clears my thoughts and breathes for me. Its benevolence is unlimited. It shepherds waves to shore — they collapse on the narrow beach with rhythmic puffs — and it eases my pain. I open my eyes long enough to see pine needles shining like threads of spun gold. A kingfisher flies from a branch with a rattling cry and disappears.

AT THE START of his book *A Story Like the Wind*, the South African writer Laurens van der Post quotes an imprisoned Bushman who is waiting in his cell for the wind to bring him news of his people. "I do merely listen," the Bushman says, "watching for a story which I want to hear; that it may float into my ear." The Polynesian peoples, too, believed that the wind carried tales, especially about the natural deities who ruled the island world. Van der Post uses the example of the Bushman to show that stories are one of life's essential elements. The "living spirit," he writes, "needs the story for its survival and renewal." I think of times when I have been rejuvenated by narratives of different sorts — a piece of fiction, a film or parable or poem, a tale told by a friend — and heartily agree.

In our myth-starved world, wind no longer brings us stories. Instead it carries poisons, cold, destruction, and fear — but it also carries fresh air, coolness, healing, and hope. It bears seeds and migrating hawks, sand and the spray of oceans. In my mind I hold an image of wind as an organic force that binds all of humankind together. I hold, as well, a hope that my vision is shared.

When I think about my own wind, the vigorous breeze of the Outer Banks, it is often as a dragon, a breathing being, predictable to a point but, like any animal, given to quirks of temperament. When I think of the global winds, the bands of air that surround the whole earth, it is as something more complex, yet at the same time singular and whole. The wind is a membrane that pulses and shifts and fastens us inside a life-sustaining vessel. The wind is a pair of hands that encircles us, massages us, comforts us, and holds us tightly to our home. We are a people doubly sealed: by the motion-filled air that hides us so well from the dark void of outer space and by the Hebrew concept of *ruwach,* by the Greek *pneuma.* By the Navajo *nilch'i,* the Iroquois *Gäoh,* the Arabic *ruh.* By wind, breath, spirit. It is all, every bit of it, the same.

Sources

In the course of my writing I drew from the work of many fine researchers. What follows is a list of the sources on which I depended most heavily.

2. Creation's First Food

pp. 13–14. I referred to the 1976 edition of *Strong's Exhaustive Concordance of the Bible* (Grand Rapids, Mich.: Grand Rapids Guardian Press). Historical information about wind in the Holy Land was taken from the *Anchor Bible Dictionary,* vol. 5 (New York: Doubleday, 1992). I also consulted *Harper's Bible Dictionary* (San Francisco: Harper San Francisco, 1985).

pp. 20–21. The version of the *Enuma elish* I used was published in *Ancient Near Eastern Texts Relating to the Old Testament,* 3rd ed., edited by James B. Pritchard (Princeton: Princeton University Press, 1969). My thanks to James L. Crenshaw of Duke University for alerting me to the similarity between the Babylonian and Judaic creation tales.

pp. 22–26. I consulted several dozen books on the religions of the world to find the references to wind contained in this chapter. A few of my sources include Xan Fielding's *Aeolus Displayed* (Francestown, N.H.: Typographeum, 1991), for the story about the Dzungarian Gate; *Primitive Culture,* by Edward B. Tylor (New York: Brentano, 1924); *The Eskimo about Bering Strait,* by Edward William Nelson (1899; reprint, Washington: Smithsonian Institution Press, 1983); Mircea Eliade's *Shamanism: Archaic Techniques of Ecstasy,* translated by Willard R. Trask (New York: Pantheon Books, 1964); and Joseph Campbell's *The Masks of God: Primitive Mythologies* (New York: Viking Press, 1969). Lyall Watson's book *Heaven's Breath: A Natural History of the Wind* (New York: William Morrow, 1984) led me to many of the myths and sources cited in this chapter.

p. 26. James Kale McNeley's *Holy Wind in Navajo Philosophy* was published in 1981 (Tucson: University of Arizona Press).

p. 30. Early Greek theories on the causes of wind can be found in *The Encyclopedia of Philosophy* (New York: Macmillan, 1967). The history of wind science is also chronicled in *Wind in Architectural and Environmental Design,* by Michele Melaragno (New York: Van Nostrand Reinhold, 1982).

p. 31. Hildegard of Bingen's eloquent writings on wind have been preserved and celebrated by Matthew Fox in *Illuminations of Hildegard of Bingen* (Sante Fe: Bear & Company, 1985).

pp. 33–35. Myths about wizardry and wind can be found in abundance in *Raising the Wind: The Legends of Lapland and Finland Wizards in Literature,* by Ernest J. Moyne (Newark, Del.: University of Delaware Press, 1981). Richard M. Dorson's *Buying the Wind* was published in 1964 (Chicago: University of Chicago Press).

p. 36. The Palestinian myth of a vengeful wind comes from David Grossman's *The Yellow Wind,* translated by Haim Watzman (New York: Farrar, Straus & Giroux, 1988). The story of the maiden Mayahuel is found in *Mexican and Central American Mythology,* by Irene Nicholson (New York: Peter Bedrick Books, 1967).

3. Guiding the Hand of History

For an engaging discussion of how and where the winds blow, and why, see *USA Today's The Weather Book,* by Jack Williams (New York: Vintage Books, 1992).

p. 41. My information on prehistoric winds comes from personal communication with Robert H. Dott, Jr., a climatologist at the University of Wisconsin in Madison, and from Dott's paper "Paleolatitude and Paleoclimate," *Wisconsin Academy of Sciences, Arts, and Letters* 67 (1979).

p. 42. I found a discussion about the prehistoric winds of the southern hemisphere in David Campbell's *The Crystal Desert: Summers in Antarctica* (Boston: Houghton Mifflin, 1992). For information on how changed wind patterns may have helped foster the great diversity of species in the tropics, Dr. Campbell kindly led me to *Biogeography and Quarternary History in Tropical America,* edited by T. C. Whitmore and G. T. Prance (Oxford, Eng.: Clarendon Press, 1987).

pp. 43–47. Accounts of how changes in atmospheric circulation patterns have affected living conditions in various parts of the world were taken from two books by Hubert H. Lamb: *Climate, History, and the Modern World* (London: Methuen, 1982, 1995), and *Climate: Present, Past, and Future* (London: Methuen, 1972); and from numerous writings by Reid A. Bryson, especially *Climates of Hunger: Mankind and the World's Changing Weather,* written with Thomas J. Murray (Madison: University of Wisconsin Press, 1977), and "Chinook Climates and Plains Peoples," *Great Plains Quarterly* 1, no. 1 (Winter 1981): 5–15.

p. 48. My version of the story of England's Lost Colony in America was drawn from David Stick's *The Outer Banks of North Carolina* (Chapel Hill: University of North Carolina Press, 1958).

p. 49. Information on the early history of sailing was taken from *The New Encyclopaedia Britannica,* 15th ed. (Chicago: Encyclopedia Britannica, 1991), and from Thor Heyerdahl's *Early Man and the Ocean* (London: George Allen & Unwin, 1978).

p. 50. Much of my material on the exploration of the world by sailing ship was drawn from *To the Ends of the Earth: The Great Travel and Trade Routes of Human History,* by Irene M. Franck and David M. Brownstone (New York: Facts on File Books, 1984). Information on Chinese exploration and trade came from *When China Ruled the Seas: The Treasure Fleet of the Dragon Throne, 1405–1433,* by Louise Levathes (New York: Simon & Schuster, 1994).

p. 52. Evidence disputing Heyerdahl's theories of Polynesian expansion can be found in many sources, including "Voyaging," by Ben R. Finney, in *The Prehistory of Polynesia,* edited by Jesse D. Jennings (Cambridge, Mass.: Harvard University Press, 1979).

p. 53. David Lewis's book *We the Navigators: The Ancient Art of Landfinding in the Pacific* was published in 1972 (Canberra: Australian National University Press).

p. 54. For a detailed and fascinating account of the Greeks' routing of the Persian navy, see Peter Green's *Xerxes at Salamis* (New York: Praeger, 1970).

p. 57. Descriptions of the North African winds encountered by forces in World War II can be found in *The War in the Desert* (New York: Time Life Books, 1972). The information on the botched American invasion of Iran came from various newsmagazines of the period and from Gary Sick's book *All Fall Down: America's Tragic Encounter with Iran* (New York: Random House, 1985).

pp. 59–63. For information on the development of wind science, I referred to a number of publications, including the *Encyclopaedia Britannica,* as cited above; *Meteorology: A Historical Survey,* vol. 1, by A. Kh. Khrgian (Leningrad: Gimiz, 1959); *A Popular Treatise on the Winds,* by William Ferrel (New York: John Wiley & Sons, 1889); *The Thermal Theory of Cyclones: A History of Meteorological Thought in the Nineteenth Century,* by Gisela Kutzbach (New York: American Meteorological Society, 1979); and *Appropriating the Weather: Vilhelm Bjerknes and the Construction of a Modern Meteorology,* by Robert Marc Friedman (Ithaca, N.Y.: Cornell University Press, 1979).

p. 64. Descriptions of the attempts by the Japanese to float bombs to America by balloon can be found in Elmer R. Reiter's "Rapid Rivers of Air," *Natural History* 84, no. 3 (March 1975): 46–51; and in John McPhee's "Annals of Crime: The Gravel Page," *New Yorker,* January 26, 1996.

p. 65. For an interesting description of clouds and cloud watching, see Fred Hapgood's "'Up in the Sky There's a Good Time to Be Had Everywhere, Always, for Free,'" *Smithsonian* 25 (April 1994): 36–43.

p. 69. The fascinating work on hurricane prediction is described in several papers by William M. Gray and Christopher W. Landsea, including a good popular account in Landsea, Gray, Paul W. Mielke, Jr., and Kenneth J. Berry, "Seasonal Forecasting of Atlantic Hurricane Activity," *Weather* 49 (August 1994). For a more technical account, see "Predicting Atlantic Seasonal Hurricane Activity 6–11 Months in Advance," by the same authors, *Weather and Forecasting* 7, no. 3 (September 1992). My information was augmented by conversations with a student of Gray's, John A. Knaff.

4. Masters of the Breeze

pp. 72–81. A wealth of scientific papers and books has been published on the way birds use wind during migration. I relied most heavily on articles by Sidney A. Gauthreaux, Jr., including "The Flight Behavior of Migrating Birds in Changing Wind Fields: Radar and Visual Analyses," *American Zoology* 31 (1991): 187–204; *Animal Migration, Orientation, and Navigation,* edited by Gauthreaux (New York: Academic Press, 1980); articles by W. John Richardson, including "Timing and Amount of Bird Migration in Relation to Weather: A Review," *Oikos* 30 (1978): 224–72; and "Wind and Orientation of Migrating Birds: A Review," in *Orientation in Birds,* edited by P. Berthold (Basel: Birkhauser Verlag, 1991). Other informative publications include Paul Kerlinger's *Flight Strategies of Migrating Hawks* (Chicago: University of Chicago Press, 1989); Norman Elkins's *Weather and Bird Behavior* (New York: Academic Press, 1988); and Jerry A. Waldvogel's "Olfactory Orientation by Birds," in *Current Ornithology,* edited by Dennis M. Power (New York: Plenum Press, 1989).

p. 82. The edition I used of J. Henri Fabre's *The Life of the Spider,* first published in 1913, was translated by Alexander Teixeira de Mattos (New York: Horizon Press, 1971).

p. 84. The account of aphids finding potted alfalfa plants in the middle of the Mojave Desert is by R. C. Dickson and is titled "Aphid Dispersal over Southern California Deserts," *Annals of the Entomological Society of America* 52 (1959): 368–72.

p. 84–85. Perry A. Glick describes his work in *The Distribution of Insects, Spiders, and Mites in the Air,* Technical Bulletin No. 673 (Washington: U.S.

Department of Agriculture, May 1939); and *Collecting Insects by Airplane in South Texas,* Technical Bulletin No. 1158 (Washington: U.S. Department of Agriculture, 1957).

pp. 86–88. For information on pheromone dispersal, I referred to "Odor Plumes and How Insects Use Them," by John Murlis, Joseph S. Elkington, and Ring T. Carde, *American Review of Entomology* 37 (1992): 505–32; and to articles by Thomas C. Baker and Neil J. Vickers, including "Reiterative Responses to Single Strands of Odor Promote Sustained Upwind Flight and Odor Source Location by Moths," *Proceedings of the National Academy of Sciences USA: Neurobiology* 91 (June 1994): 575–660. My information was augmented by conversations with Dr. Allard Cossé of Iowa State University.

pp. 86–90. Lawrence W. Swan's fascinating work on the aeolian zone is discussed in "Aeolian Zone," *Science* 140, no. 5 (April 1963): 7778, and in "The Aeolian Biome: Ecosystems of the Earth's Extremes," *BioScience* 42, no. 4 (April 1992): 262–70.

pp. 92–93. Steven Vogel's research on animals and their uses of wind has appeared in many places. See especially his book *Life in Moving Fluids: The Physical Biology of Flow* (Princeton: Princeton University Press, 1981; rev. ed., 1996), and "Organisms That Capture Currents," *Scientific American* 239, no. 2 (August 1978): 128–35.

5. The Voices of Trees

pp. 96, 100. A number of papers on the damage caused by Hurricane Andrew in south Florida ecosystems appear in the *Journal of Coastal Research,* special issue no. 21 (Spring 1995). See especially "Effects of Hurricane Andrew on Coastal and Interior Forests of Southern Florida: Overview and Synthesis," by T. V. Armentano and others. I drew background information on pine rocklands and the Miami Rock Ridge from Joe Maguire's *Restoration Plan for Dade County's Pine Rockland Forests Following Hurricane Andrew* (Dade County Department of Environmental Resource Management, August 1995).

pp. 98–99. Much of the general information on plants in this chapter was taken from *Biology of Plants,* 4th ed., by Peter H. Raven, Ray F. Evert, and Susan E. Eichhorn (New York: Worth Publishers, 1986). The quote from Thoreau was taken from "The Dispersion of Seeds," an unfinished manuscript of Thoreau's that was pieced together by Bradley P. Dean and published in *Faith in a Seed* (Washington: Island Press, 1993).

pp. 101–2. Robert Doren of Everglades National Park kindly helped me with information about the restoration of the Hole in the Donut, and he offered a brief explanation of why slash pines within the park survived the hurricane while those outside did not. For more precise information on the restoration work, see "Restoration of Former Wetlands within the Hole-in-the-Donut in Everglades National Park," by Doren and others, in *Proceedings of the Seventeenth Annual Conference on Wetlands Restoration and Creation,* edited by Frederick J. Webb (Plant City, Fla.: Hillsborough Community College, 1990).

p. 108. An interesting description of microclimates is found in Jay Stuller's "If You Don't Like the Climate, Just Walk a Few Yards," *Smithsonian* 20, no. 9 (December 1995).

pp. 108–9. Information on turbulence and its circulation through plant canopies was drawn from Paul Sheldon Conklin's Ph.D. dissertation, "Turbulent Wind and Pressure in a Mature Hardwood Canopy," Duke University, 1994, and from several of the papers in *Wind in Trees,* edited by M. P. Coutts and J. Grace (Cambridge, Eng.: Cambridge University Press, 1995). An excellent description of the balancing act that trees must perform when the wind kicks up can be found in Steven Vogel's popular book *Life in Moving Fluids: The Physical Biology of Flow* (Princeton: Princeton University Press, 1981; rev. ed., 1996). See also *Plant Responses to Wind,* by J. Grace (London: Academic Press, 1977).

p. 109. The description of ways in which plants pollinate themselves in windless rain forests was taken from *Tropical Nature: Life and Death in the Rain Forests of Central and South America,* by Adrian Forsyth and Ken Miyata (New York: Scribner's, 1984).

p. 110. The quote from Xan Fielding was taken from his book *Aeolus Displayed* (Francestown, N.H.: Typographeum, 1991).

pp. 111–13. Steve Vogel has published many papers and articles on the effects of wind on trees and plants. His article in *Natural History* appeared in September 1993. For a more scientific treatment, see "Drag and Reconfiguration of Broad Leaves in High Winds," *Journal of Experimental Botany* 40, no. 217 (August 1989): 941–48. Other interesting studies by Vogel include "Convective Cooling at Low Airspeeds and the Shapes of Broad Leaves," *Journal of Experimental Botany* 21, no. 66 (February 1970): 91101; and "Twist-to-Bend Ratios of Woody Structures," *Journal of Experimental Botony* 46, no. 289 (August 1995): 981–85.

p. 119. Ann H. Zwinger's book *Land above the Trees* was coauthored with Beatrice E. Willard (New York: Harper & Row, 1972). I drew much information about mountain plants from this fine volume.

6. The Waltzing of Wind and Sand

There are a number of good texts on the cross stratification of aeolian bedforms. For a basic understanding of the subject, I depended on the book

Aeolian Sand and Sand Dunes, by Kenneth Pye and Haim Tsoar (London: Unwin Hyman, 1990).

p. 128. The edition of Ralph A. Bagnold's *The Physics of Blown Sand and Desert Dunes* that I used was issued in 1965 (London: Methuen & Co.; originally published in 1941). I looked also at Bagnold's *Libyan Sands: Travel in a Dead World* (London: Hodder & Stoughton, 1935).

p. 130. A number of American investigators have done exemplary work in interpreting ancient aeolian bedforms. A seminal paper, "Basic Types of Stratification in Small Eolian Dunes," was published by Ralph E. Hunter of the U.S. Geological Survey, Menlo Park, California, in *Sedimentology* 24 (1977). Hunter went on to publish more on dune strata with David M. Rubin, also of the Geological Survey in Menlo Park. Rubin has developed detailed computer models on bedforms of aeolian structures. Other investigators of note who have worked extensively on the aeolian bedforms of the Colorado Plateau include Ronald C. Blakey of Northern Arizona University in Flagstaff and Fred Peterson of the U.S. Geological Survey in Denver.

pp. 137–40. For a summary of the model that grew out of the interpretive work of Gary Kocurek and Karen Havholm, see "Eolian Sequence Stratigraphy — A Conceptual Framework," in *Siliciclastic Sequence Stratigraphy,* edited by P. Weimer and H. W. Posamentier, American Association of Petroleum Geologists Memoir no. 58 (1993). The interpretation of the bedforms that make up Julia's Knob is detailed in Gary Kocurek, Julia Knight, and Karen Havholm, "Outcrop and Semi-Regional Three-Dimensional Architecture and Reconstruction of a Portion of the Eolian Page Sandstone (Jurassic)," in *The Three-Dimensional Facies Architecture of Terrigenous Clastic Sediments and Its Implications for Hydrocarbon Discovery and Recovery,* edited by Andrew D. Miall and Noel Tyler, vol. 3 of *SEPM* (Society for Sedimentary Geology) *Concepts in Sedimentology and Paleontology* (1991), pp. 25–43.

7. The Encircling Seas

For a good overview of oceanographic processes and wind, see *An Introduction to the World's Oceans,* by Alyn C. Duxbury and Alison Duxbury (Reading, Mass.: Addison-Wesley, 1984).

p. 147. Information on the effects of the west (or north) wall of the Gulf Stream on local winds came from interviews with personnel at the U.S. Naval Atlantic Meteorology and Oceanography Center in Norfolk and from "Up Against the Wall," in the navy magazine *Surface Warfare* (December 1978).

pp. 149–50. Doron Nof and Nathan Paldor's articles about the parting of the Gulf of Suez are as follows: "Are There Oceanographic Explanations for the Israelites' Crossing of the Red Sea?" *Bulletin of the American Meteorological Society* 73, no. 3 (March 1992): 305–14, and "Statistics of Wind over the Red Sea with Application to the Exodus Question," *Journal of Applied Meteorology* 33, no. 8 (August 1994): 1017–25.

p.151. The study on meteorological bombs by researchers at the Massachusetts Institute of Technology is described in "Synoptic-Dynamic Climatology of the 'Bomb,'" *Monthly Weather Review of the American Meteorological Society* 108 (October 1980): 1589–1606.

p. 152, 154–58. I am indebted to Vincent J. Cardone of Oceanweather, Inc., for his help in understanding wind and wave building. Dr. Cardone and several other researchers have published a number of fascinating papers on bombs and wave hindcasting, in particular "Evaluation of Contemporary Ocean Wave Models in Rare Extreme Events: The 'Halloween Storm' of October 1991 and the 'Storm of the Century' of March 1993," *Atmosphere and Ocean Technology* 13, no. 1 (February 1996): 198–230.

pp. 153–54. Sebastian Junger's *The Perfect Storm: A True Story of Men against the Sea* was published in 1997 (New York: W. W. Norton).

pp. 154–56. V. R. Swail of Canada's Climate and Atmospheric Research Directorate in Ontario was tremendously helpful in sending unpublished posters and bulletins on Hurricane Luis.

pp. 158–59. Work on Portuguese men-of-war and their sailing direction was conducted primarily by A. H. Woodcock of the Woods Hole Oceanographic Institution. See, for example, "Dimorphism in the Portuguese Man-of-War," *Nature* (August 4, 1956): 253–55.

p. 159. A description of the experiments on loggerhead turtle navigation can be found in "How Sea Turtles Navigate," *Scientific American* (January 1992): 100106.

p. 160. The study of seabird feeding and wind-dispersed odors is described in an article by Les Line in the *New York Times*, August 29, 1995.

pp. 160–61. I drew information on albatrosses from the *Audubon Society Encyclopaedia of North American Birds*, by John K. Terres (New York: Knopf, 1982), and David G. Campbell's *The Crystal Desert: Summers in Antarctica* (Boston: Houghton Mifflin, 1992).

p. 162. Steven J. Lentz of the Woods Hole Institution kindly provided me with several papers on upwelling off the California coast and elsewhere. For an overview, see his "U.S. Contributions to the Physical Oceanography of Continental Shelves in the Early 1990s," *U.S. National Report to International Union of Geodesy and Geophysics 1991–1994, Review of Geophysics, Supplement* (July 1995), pp. 1225–36.

pp. 164–65. A popular account of Cheryl Ann Butman's experiments can be found in "Measuring Diversity of Planktonic Larvae," *Oceanus* 39, no. 1

(Spring–Summer 1996): 12. A more technical, if slightly dated, treatment appears in "CoOP: Coastal Ocean Processes Study," *Sea Technology* (January 1994): 44–49.

pp. 165–67. Studies by Curtis C. Ebbesmeyer and W. James Ingraham, Jr., are described in "Shoe Spill in the North Pacific," *Eos, Transactions, American Geophysical Union* 73, no. 34 (August 25, 1992): 361, 365, and "Pacific Toy Spill Fuels Ocean Current Pathway Research," *Eos* 75, no. 37 (September 13, 1994): 425, 427, 428.

pp. 167–68. Leonard J. Pietrafesa kindly sent me numerous studies on the transport of fish larvae, including two recently completed manuscripts, "The Hydrodynamics and Volumetric Flux through Oregon Inlet, NC," by C. Reid Nichols and Pietrafesa, June 1997, accepted by *Journal of Coastal Research,* and "Physical Oceanographic Processes Affecting Larval Recruitment of Estuarine Dependent Finfish via Beaufort Inlet," by D. G. Logan, Pietrafesa, and others, unpublished report, March 1997.

8. Of Body and Mind

Throughout this chapter I write of various regional winds that are known as either healing or poisonous. Much of my information was taken from Slater Brown's *World of the Wind* (New York: Bobbs-Merrill, 1961); Stephen Rosen's *Weathering* (New York: M. Evans, 1979); and Lyall Watson's *Heaven's Breath* (New York: William Morrow, 1984).

pp. 172, 175–76. The description of medical climatology as an inexact science comes from Laurence S. Kalkstein and Robert E. Davis, "Weather and Human Mortality: An Evaluation of Demographic and Interregional Responses in the United States," *Annals of the Association of American Geographers* 79, no. 1 (1989): 44–64.

p. 173. Information about the physical effects of wind on humans is taken from several articles in *Medical Climatology,* edited by Sidney Licht (New Haven: Elizabeth Licht, 1964). The most informative chapter in this collection of medical essays is "Effects of Wind on Man," by Jozef Jankowiak.

p. 175. The study on how wind agitates young children is described in an article by Eva L. Essa, Jeanne M. Hilton, and Colleen I. Murphy, "The Relationship between Weather and Preschoolers' Behavior," *Children's Environmental Quarterly* 7, no. 3 (1990): 32–36.

pp. 176–77. The study of arthritic patients in Chicago and North Dakota was by Joyce M. Laborde, William A. Dando, and Marjorie J. Powers, "Influence of Weather on Osteoarthritics," *Social Science and Medicine* 23, no. 6 (1986): 549–54.

p. 178. I found C. L. Edson's poem "The Prairie Pioneers" in *The Kansas*

Experience in Poetry (Lawrence: University of Kansas Press, 1986). My thanks to Thomas Fox Averill, who sent me this and several other poems.

pp. 179–80. Dorothy Scarborough's novel *The Wind* was published, at first anonymously and amid great controversy, in 1925 (New York: Harper & Brothers). The edition I found was issued in 1979 (Austin: University of Texas Press).

p. 181. John Calderazzo's comments come from his wonderful essay "Sailing Through the Night," *Orion* (Winter 1995).

pp. 182–83. Thoughts on the Los Angeles wind by Joan Didion were taken from "Los Angeles Notebook," in *Slouching Towards Bethlehem* (New York: Farrar, Straus & Giroux, 1990).

pp. 187, 200–201. Information on the structure of hurricanes and tornadoes was taken from interviews with scientists and from *USA Today's The Weather Book,* by Jack Williams (New York: Vintage Publishers, 1992). This is an entertaining source book for anyone interested in weather.

p. 191. Mary Swander's *Out of This World: A Journal of Healing* was published in 1995 (New York: Viking Penguin).

p. 205. The attempts by scientists to destroy two hurricanes is chronicled by R. H. Simpson and Joanne S. Malkus in "Experiments in Hurricane Modification," *Scientific American* (December 1964), reprinted in *The Enigma of Weather* (Scientific American, 1994).

9. *Tapping the Flow*

p. 207. The story of Minamoto-no-Tametomo was taken from *The Penguin Book of Kites,* by David Pelham (New York: Penguin Books, 1976).

p. 213. The history of Aeolian harps came from several sources. I drew most heavily from the liner notes for Roger Winfield's compact disk *Windsongs* (Wotton-Under-Edge, Eng.: Saydisc Records, 1991).

p. 214. Information about smelting furnaces in Sri Lanka came from an article by John Noble Wilford in the *New York Times,* February 6, 1996.

p. 214–16. My main sources for windmill history were *Power from Wind: A History of Windmill Technology,* by Richard L. Hills (Cambridge, Eng.: Cambridge University Press, 1994); and *Medieval Religion and Technology: Collected Essays,* by Lynn White, Jr. (Berkeley: University of California Press, 1978).

p. 217. The edition of Antoine de Saint-Exupéry's *Wind, Sand and Stars* that I consulted was published in 1940 (New York: Harcourt, Brace & World.)

p. 218. I read of problems with clear-air turbulence on flights over the Colorado Rockies in the *Wall Street Journal,* March 28, 1997, and over the North Pacific in the *Norfolk Virginian Pilot,* December 29, 1997. Information about wind-shear warning devices came from an article in the *Virginian Pilot,* December 3, 1994, and from conversations with engineers at NASA's Langley Research Center.

pp. 221–26. Information on the *Dymocks Flyer* balloon and space capsule came from team publications and interviews. Accounts of attempts by other teams to fly around the world were taken from newspaper articles and interviews. The description of wind flow in the stratosphere was taken from conversations with researchers at the National Balloon Facility in Palestine, Texas.

p. 227. For an overview of the history of wind power, see Paul Gipe's *Wind Energy Comes of Age* (New York: John Wiley & Sons, 1996).

pp. 227–34. The statistics on wind energy production and potential, both in the United States and overseas, was found in materials kindly provided by the library at the U.S. Department of Energy's National Wind Technology Center in Golden, Colorado, and by employees of the American Wind Energy Association and the California Public Utility Commission. I am indebted to Nancy Rader of the Berkeley office of AWEA and to Stuart Chaitkin of CPUC. The National Wind Coordinating Committee has a number of informative position papers detailing the advantages and disadvantages of wind power. For an interesting discussion of the renewables portfolio standard, see "Efficiency and Sustainability in Restructured Electric Markets: The Renewables Portfolio Standard," by Nancy A. Radar and Richard B. Norgaard, *Electricity Journal* (July 1996): 37–49.

10. Taunting the Wind

p. 241. The description of airflow in the Pantheon and the work of Vitruvius came from Michele Melaragno's *Wind in Architectural and Environmental Design* (New York: Van Nostrand Reinhold, 1982).

p. 242. For an interesting discussion of the use of airflow in the desert, see "Passive Cooling Systems in Iranian Architecture," *Scientific American* 238, no. 2 (February 1978). Sydney A. Bagg's unpublished study, "The Dugout Dwellings of an Outback Opal Mining Town in Australia," was given to me by Steven Vogel of Duke University.

p. 243. Information on the problems with the John Hancock Tower in Boston was taken from articles in the *Boston Globe* that appeared on April 9, 1988, and March 3, 1995.

pp. 243–44. Accounts of bridge collapses in the nineteenth and twentieth centuries can be found in Melaragno's book, cited above, and in Henry Petroski's *Engineers of Dreams: Great Bridge Builders and the Spanning of America* (New York: Knopf, 1995).

p. 244. Pitcher Stu Miller's travails during the 1961 All Star Game are recounted in "Taming the Winds," by Jack E. Cermak, in *Science Year: The World Book Science Annual* (New York: Field Enterprises Educational Corporation, 1977). A detailed history of the science of wind engineering can be found in Jack Cermak's "Applications of Fluid Mechanics to Wind Engineering — A Freeman Scholar Lecture," *Journal of Fluids Engineering* 97 (March 1975).

p. 249. For a popular discussion of wind flow around skyscrapers, including a series of very clear diagrams, see Richard Rush's "Technics: Structuring Tall Buildings," *Progressive Architecture* (December 1980).

p. 263. Michael Specter's article on the aftereffects of the Chernobyl nuclear accident appeared in the *New York Times* on March 31, 1996.

p. 264. For a history of early air pollution problems, see Melaragno's book, cited above, and Richard A. Prindle's "Air Pollution and Community Health," *Medical Climatology*, edited by Sidney Licht (New Haven: Elizabeth Licht, 1964). Many sources describe the early detection of acid rain. For a well-researched, well-written account, see Louise B. Young's *Sowing the Wind: Reflections on the Earth's Atmosphere* (New York: Prentice Hall, 1990).

p. 266. Reports of acid rain damage to the African rain forest can be found in Louise B. Young's *Sowing the Wind*, listed above.

Epilogue: A World of Great Wind

p. 269. The edition I used of Laurens van der Post's *A Story Like the Wind* was published in 1972 (New York: William Morrow).

Index